Simplifying Hybrid Cloud Adoption with AWS

Realize edge computing and build compelling hybrid solutions on premises with AWS Outposts

Frankie Costa Negro

BIRMINGHAM—MUMBAI

Simplifying Hybrid Cloud Adoption with AWS

Group Product Manager: Rahul Nair
Publishing Product Manager: Meeta Rajani
Senior Editor: Arun Nadar
Technical Editor: Shruthi Shetty
Copy Editor: Safis Editing
Project Coordinator: Ashwin Kharwa
Proofreader: Safis Editing
Indexer: Rekha Nair
Production Designer: Roshan Kawale
Marketing Coordinator: Nimisha Dua

First published: November 2022

Production reference: 1041122

Published by Packt Publishing Ltd.
Livery Place
35 Livery Street
Birmingham
B3 2PB, UK.

ISBN 978-1-80323-175-4

www.packt.com

To all AWS friends and the inspiring leaders Andy Jassy, Werner Vogels, James Hamilton, Jeff Barr, Peter DeSantis, and Anthony Liguori.

Contributors

About the author

Frankie Costa Negro is a technology wanderer. He began in IT as a FoxPro / dBase developer in 1994. Quite a lot has happened since then: heroic days learning Microsoft Office **Visual Basic for Applications** (**VBA**). Creating macros with Microsoft Excel 5.0, and transitioning to Visual Basic 3.0. His developer career ended with Visual Basic 6.0 when he was invited to be a sysadmin in 1997. A lot of water has gone under the bridge and his on-premises stint ended in 2018 with an invitation to join AWS as a Cloud Support Engineer. Since then, it has all been hard work while having fun and making history. He came full circle by joining the AWS Outposts team to flex his on-premises muscles once again. When he is not working for AWS, he is working accompanied by music.

I want to thank my wonderful wife and soulmate, Claudia, for the relentless cheering-on and support that fueled my will to persevere throughout this journey. Heartfelt gratitude to the whole Packt editing team for being the North Star for this first-time book author. I want to acknowledge my beloved daughter, Giovanna, my inspiration for fulfilling the famous quote that tells us three things every human being should do during their time on Earth: "Plant a tree, have a child, and write a book." Finally, I acknowledge each and every person that meets me with a smile and affection: you make me believe there is meaning and hope in this life so that it is worth living.

About the reviewer

Harsha Sanku is a Senior Partner Solutions Architect with AWS specializing in VMware Cloud on AWS, AWS Outposts, and AWS Networking. He has worked with customers and strategic partners, helping them design and build secure, resilient, and scalable environments using hybrid cloud services that deliver a consistent experience from the cloud to on-premises and at the edge. He also works with various partner engineering and service teams to build roadmaps and drive service enhancements. He has been working in the IT industry in multiple roles since 2006.

Table of Contents

Part 2: Security, Monitoring, and Maintenance

6

Monitoring Outposts   171

Part 3: Maintenance, Architecture References, and Additional Information

7

Outposts Maintenance   207

8

Architecture References   215

Index   223

Other Books You May Enjoy   232

Preface

AWS Outposts is the next wave in technology, bringing cloud infrastructure to customer premises. Understanding how to leverage this product to address hybrid scenarios is key to unleashing the full potential of the service and paving the way for your cloud journey. This book provides extensive coverage of everything Outposts, from zero to hero. You will begin your journey learning about the concepts of Hybrid Edge, transitioning to understand what AWS Outposts is and how it fits in this space alongside its common use cases. Next, you will have a tour that unveils its components, connections, and capabilities. With this in-depth knowledge, you will discover the processes and steps involved in having Outposts delivered and ready to use. Then, it is finally time to see your Outposts rack in action and leverage AWS services running in the rack. Security is a fundamental discipline and you will get detailed explanations about the security features in Outposts and thoroughly understand how to monitor and maintain Outposts. Your journey will conclude with a roadmap to raise your Outposts knowledge to a professional level, becoming a true Outposts hero!

Who this book is for

This book is intended for data center architects interested in understanding, designing, architecting, and building solutions with AWS Outposts, leveraging the product's capabilities and best practices. DevOps professionals will understand how to build compelling hybrid solutions and applications using the same tools and technologies available in AWS regions. Business and project managers working on projects with AWS Outposts will also benefit from this book at a higher level, without the need to get into very technical details.

What this book covers

Chapter 1, *Introduction to AWS Outposts*, opens with an elaboration of the section or area called the *Edge* in the IT realm because that is typically where Outposts lives, then transitions to cover what Outposts is, exploring some of its key concepts and terminology, wrapping up with some valid use cases this product is best suited for.

Chapter 2, *AWS Outposts Anatomy*, pops the hood to show Outposts components, how they interact with each other as well as with the environment and hooks within the facility where Outposts lives, how it operates with the underlying networks, and how it exposes its resources. Subsequently, we talk about the AWS services that can be leveraged in Outposts and how it connects to an AWS Region.

Chapter 3, Pricing, Ordering, and Installation, walks you through the ordering process step by step and talks about configuration options, pricing, and what happens after an order is completed so you can finally get your Outposts rack delivered to your facility.

Chapter 4, Operations and Working with Outposts Resources, talks about the setup procedure and some requirements. Finally, let's start the Outposts engine and shows how to effectively use the product and begin to use the building blocks and the capabilities of AWS services it is capable of running to craft your hybrid application or use case.

Chapter 5, Security Aspects in Outposts, leverages the breadth and depth of AWS security services and capabilities to meet the most stringent security requirements. As it takes AWS technology outside the confines of AWS regions, controls and safeguards were added to the product and the Shared Security Model was changed a bit to assign an extra layer of responsibility to the customer.

Chapter 6, Monitoring Outposts, focuses on the metrics and visibility you can get from Outposts. To understand what is going on inside your system you need to translate collected data into information so you can take actions to control undesired events and behaviors. Furthermore, you can use that information to improve your architecture and operations.

Chapter 7, Outposts Maintenance, describes the shared responsibility model, a security and compliance framework that outlines the responsibilities of AWS and the customer. AWS is responsible for the hardware and software that runs AWS services and that includes maintenance tasks on AWS Outposts. These activities are highlighted alongside useful hints on how to troubleshoot connectivity issues.

Chapter 8, Architecture References, concludes our journey with cornerstone information about how to best leverage the product architecture to enable you to create compelling hybrid solutions using AWS Outposts.

To get the most out of this book

This book assumes that technical professionals possess strong foundations in data center technologies, servers, networking, and storage architectures and concepts and knowledge at the *AWS Architect Associate Certification* level. As for business professionals, a high-level understanding of technical terms and the building blocks of an IT infrastructure is assumed.

Software/hardware covered in the book	Operating system requirements
AWS CLI	Linux
AWS CloudShell	Windows

If you are using the digital version of this book, we advise you to type the code yourself or access the code from the book's GitHub repository (a link is available in the next section). Doing so will help you avoid any potential errors related to the copying and pasting of code.

Check the GitHub repository to find a collection of URLs to take your knowledge about Outposts to the professional level.

Download the example code files

You can download the example code files for this book from GitHub at `https://github.com/PacktPublishing/Simplifying-Hybrid-Cloud-Adoption-with-AWS`. If there's an update to the code, it will be updated in the GitHub repository.

We also have other code bundles from our rich catalog of books and videos available at `https://github.com/PacktPublishing/`. Check them out!

Download the color images

We also provide a PDF file that has color images of the screenshots and diagrams used in this book. You can download it here: `https://packt.link/amQG4`.

Conventions used

There are a number of text conventions used throughout this book.

`Code in text`: Indicates code words in text, database table names, folder names, filenames, file extensions, pathnames, dummy URLs, user input, and Twitter handles. Here is an example: "To create the remaining two subnets for this CIDR block, repeat the command adjusting the `--availability-zone-id` parameter accordingly, and also make note of the new subnet IDs."

A block of code is set as follows:

```
{
    "LocalGateways": [
        {
            "LocalGatewayId": "lgw-0cdc67d1ae6c75ff8",
            "OutpostArn": "arn:aws:outposts:us-west-
2:123456789012:outpost/op-8b286039iedad23e0",
            "OwnerId": "123456789012",
            "State": "available",
            "Tags": []
        }
    ]
}
```

Any command-line input or output is written as follows:

```
aws ec2 describe-local-gateway-virtual-interface-groups \
    --local-gateway-virtual-interface-group-ids "lgw-vif-grp-
033d2b33464749f3a" \
    --region "us-west-2";
```

Bold: Indicates a new term, an important word, or words that you see on screen. For instance, words in menus or dialog boxes appear in **bold**. Here is an example: "Select **System info** from the **Administration** panel."

> **Tips or important notes**
> Appear like this.

Get in touch

Feedback from our readers is always welcome.

General feedback: If you have questions about any aspect of this book, email us at customercare@ packtpub.com and mention the book title in the subject of your message.

Errata: Although we have taken every care to ensure the accuracy of our content, mistakes do happen. If you have found a mistake in this book, we would be grateful if you would report this to us. Please visit www.packtpub.com/support/errata and fill in the form.

Piracy: If you come across any illegal copies of our works in any form on the internet, we would be grateful if you would provide us with the location address or website name. Please contact us at copyright@packt.com with a link to the material.

If you are interested in becoming an author: If there is a topic that you have expertise in and you are interested in either writing or contributing to a book, please visit authors.packtpub.com.

Share Your Thoughts

Once you've read *Simplifying Hybrid Cloud Adoption with AWS*, we'd love to hear your thoughts! Scan the QR code below to go straight to the Amazon review page for this book and share your feedback.

https://packt.link/r/1803231750

Your review is important to us and the tech community and will help us make sure we're delivering excellent quality content.

Download a free PDF copy of this book

Thanks for purchasing this book!

Do you like to read on the go but are unable to carry your print books everywhere?

Is your eBook purchase not compatible with the device of your choice?

Don't worry, now with every Packt book you get a DRM-free PDF version of that book at no cost.

Read anywhere, any place, on any device. Search, copy, and paste code from your favorite technical books directly into your application.

The perks don't stop there, you can get exclusive access to discounts, newsletters, and great free content in your inbox daily

Follow these simple steps to get the benefits:

1. Scan the QR code or visit the link below

https://packt.link/free-ebook/9781803231754

2. Submit your proof of purchase
3. That's it! We'll send your free PDF and other benefits to your email directly

Part 1: Understanding AWS Outposts – What It Is, Its Components, and How It Works

This part sets the stage for the next parts of the book. You will understand what AWS Outposts is, the tenets that drive the product design and engineering, and valid product use cases. Furthermore, it dives deep into the innards of the product and its interactions with the infrastructure it is built upon, closing with the price perspective, how to order an Outposts rack, and what happens before it is finally activated at the operating site.

This part has the following chapters:

- *Chapter 1*, An *Introduction to AWS Outposts*
- *Chapter 2, AWS Outposts Anatomy*
- *Chapter 3, Pricing, Ordering, and Installation*

1
An Introduction to AWS Outposts

After prevailing over the initial hype cycle, the cloud has truly evolved as the de facto platform of choice to run IT services, by overcoming what seemed to be an insurmountable gap between customer premises and the cloud. There have been several attempts to bridge both these worlds.

The term *hybrid* was initially coined as a reference to solutions that purported to operate as a cloud would in the customer data center. Ever since then, several of these kinds of denominations in the Information Technology realm have evolved and consolidated into what is now referred to as the edge. **Amazon Web Services** (**AWS**) now delivers managed cloud infrastructure in the form of AWS Outposts, which lives in the very same edge.

This chapter explores the concept of the edge and then transitions into exploring the key concepts and terminology of AWS Outposts. Finally, we will wrap up with the use cases for this product.

In this chapter, you will cover the following:

- Identifying the edge space in the Information Technology domain
- Understanding the purpose of AWS Outposts
- Identifying how Outposts fits into the edge space
- Understanding what business problems Outposts solves

Defining hybrid, edge, and rugged edge IT spaces

Amazon as an enterprise has rapidly evolved from the challenges it endured that could neither have been addressed technically nor economically at the time when running large-scale applications. This fact may have led us to conclude that AWS was unlikely to develop a product that would resemble a traditional server rack you could find in any regular data center.

As with any market or industry, things change. New technologies arise, paradigms shift, and new trends pose new challenges and require new solutions. It was no different from the way enterprises consume corporate IT services. In the past couple of years, the hybrid phenomenon gained a lot of momentum to become one of the preferred ways for enterprises to run their business.

This is not strange by any means. You have the start-up sector, which is cloud-native and certainly does not see any reason to have a physical infrastructure of its own. Start-up companies only need a solid connection to the internet and personal equipment to carry out the development work and the administrative work with cloud providers.

At the other end of the spectrum, we had companies over the past few decades doing IT the traditional way, operating fully on-premises. But in recent years, the market has developed the perception that it does not need to be one way or the other.

Back then, your option to run IT infrastructure outside your own local data center relied on offerings from third-party specialized data center providers. Offerings such as hosting, location, and co-location were extremely popular at the time, and are still available today. If you could order a good leased line to connect your site(s) with the provider's site, any of the options would be available to you.

At best, you could have one of these providers supplying and managing all the necessary equipment to run your business while leveraging the OPEX financial model. Your IT team would take care of services and **Line-Of-Business** (**LOB**) applications and you would be in business. For some companies, the CAPEX model made sense as purchased equipment became assets for the company and added value to the balance sheets.

Times change and the advent of the cloud challenged the constraints and limitations of traditional data centers. Andrew Jassy, currently the president and CEO of Amazon, in an interview for the site TechCrunch (`https://techcrunch.com/2016/07/02/andy-jassys-brief-history-of-the-genesis-of-aws/`), described how AWS was conceived to be the *Operating System for the internet* at its inception, designed to reliably run applications and services at massive scale.

When AWS came to life in 2006, it wasn't too clear that it would become what it is today. From humble beginnings with no clear ambition and marketing to turning into a cloud behemoth just 15 years later, AWS and the cloud were just yet another technology trend that remained to prove reliable and solid. The early adopters pioneered the new cloud paradigm and got their feet wet with infrastructure and services that existed beyond their reach when they could not even schedule a visit to the data center.

Adopting cloud services was an exercise of a dual IT landscape design. Either one given service lived on-premises or lived in the cloud. The connection between the two was basically to exchange data for migration or backup and, eventually, very simple interaction between systems with multiple components. It was difficult to consider a three-tier architecture where one of the tiers would sit in the cloud with the other on-premises. Internet bandwidth was scarce, connections were not strongly reliable, and you often had to resort to VPNs for security because it was challenging to procure dedicated links to directly connect with cloud providers at the time.

As the cloud trend reached critical mass and established itself as a valid path, businesses faced a new reality: the cloud had to be considered within their technology plans and a thorough assessment of the IT landscape was necessary to devise a strategy that could somehow encompass cloud offerings and to a serious extent. A vague statement about the cloud just being hype was no longer acceptable to business owners – it was here to stay.

This new way of consuming corporate IT services was dubbed **hybrid cloud** and described as a combination of cloud services running alongside the traditional on-premises data center solutions. Not surprisingly, the point of view of this model was oriented from the data center out into the outside world, stretching toward the cloud, because it was primarily articulated by on-premises infrastructure providers whose vision centered around the traditional model.

The possibility of a business going all in with the cloud while shutting down all traditional data centers was somewhat far-fetched, but it was delineated as a real alternative. While it is clear that not all workloads will be a fit for the cloud and some may remain on-premises, a significant shift of IT infrastructure to the cloud can realistically be envisioned.

Further developments in this trend revealed that one piece of the puzzle was missing. If considered as a binary choice, an on-premises data center versus the cloud, any move could be a significant risk because there was no middle ground. IT teams were facing an *all-or-nothing* situation where systems with multiple components would have to be moved as a whole, likely in one go.

Evaluating how a system would perform when running on the cloud was complex because tests had to be carried out in terms of production size and capacity without close contact with all other surrounding systems and services. Even with extensible tests, a cutover date was an event of high significance, full of anxiety, and likely to have a long maintenance window. Clearly, an intermediary infrastructure bridging both worlds would be beneficial.

Initial attempts to fill this gap were made by traditional software providers, offering solutions to be run on-premises that used the type of technologies and solutions offered by cloud providers. This was the *private cloud* – one attempt to bring the cloud operational model to customer on-premises data centers. Running on their own infrastructure at their data centers or co-location sites, the promise was to leverage cloud-like services and technologies at your facility or closer to you.

It was a good approach and makes good sense. IT teams can become familiar with cloud technologies and how system operations are carried out in the cloud while relatively comfortable at home with their own equipment, learning at their own pace. As IT professionals became familiar with the cloud model, the transition to a cloud provider could be facilitated as the value and challenges became clearer.

Even with a good portion of the market leveraging the *private cloud* offering, there was still the inescapable fact that on-premises, you could not leverage the cloud-specific services and technologies. Moreover, you would never benefit from the scalability and economies of scale offered by cloud providers. It was you running cloud-like services and still managing the necessary infrastructure.

Cloud adoption has gained significant momentum in recent years and we can see now how start-up companies are said to be *born in the cloud* or *cloud natives*. These businesses would have never considered creating their products and applications using the on-premises infrastructure. Such offerings would not be possible if they were conceived within the limitations and paradigms of traditional technologies.

Systems have become increasingly complex, made up of many moving parts as opposed to the monolithic approach of yesterday. Technologies favored distributed systems and highly specialized and smaller microservices. This movement highly favored the appeal of the cloud, built on top of pay-per-use, faster innovation, elasticity, and scale. For more information, refer to this video (`https://www.youtube.com/watch?v=yMJ75k9X5_8`), *The Six Main Benefits of Cloud Computing*.

Fast forward to today and considering the latest world developments, the cloud has completely solidified its position and, to be fair, has exploded in adoption, which was significantly accelerated because of the challenges imposed by recent events such as the pandemic. The cloud model was battle-tested and made it through, to the point that it became the *de facto* standard model to be considered the foundation of technology.

While the future of the cloud seems to be clear skies, there is another fact that still holds: the vast majority of IT spending is still on traditional infrastructure and data centers. While this seems to be a wonderful opportunity to thrive in a market where the largest chunk of business is yet to be conquered, it also means that the missing key piece to act as the catalyst for the widespread adoption of the cloud is more crucial than ever.

As the next step toward blurring the boundaries between the cloud and the so-called *physical* world, the concept of a hybrid has been redefined. Hybrid is considered to be this enabler, the indistinguishable middle ground where on-premises and the cloud live together in a harmonic symbiosis where both parties benefit from each other. To amplify that notion, the term **edge** was added to the vernacular.

What we are now seeing is the original *hybrid* concept in reverse. Now, it originates in the cloud and branches out to the world in the form of edge nodes, where any given data center is considered to be one of these nodes. Effectively, the cloud aims to be everywhere, encompassing all kinds of businesses and places, powered by the recent advancements in high-speed wireless connectivity through 5G networks and IoT devices and sensors.

To make it clearer, an edge node is considered to be anywhere you could run some form of computing, be it large, small, or tiny. Naturally, a family house, a hospital, a restaurant, a crop field, an underground mine, and a cargo ship are significantly different places in nature. Suitability to accommodate electronic components and connectivity conditions change radically and the mileage of the IT equipment running will vary.

To describe these components better when deployed in harmful and aggressive environments, this space is conceptualized as the **rugged edge**, where equipment must withstand harsh usage conditions and must incorporate design characteristics and features that allow prolonged, normal operation under those circumstances. Equipment built for this purpose boasts specs that allow for severe thermal, mechanical, and environmental conditions.

Today, cloud companies are challenging themselves to create technologies that will propel the ultra-connected world where technology is pervasive, data is collected massively everywhere, and information is nearly real-time. Hybrid solutions play a fundamental role in this game, paving the way for cloud providers to extend all over the world and become *the* infrastructure, not *one* infrastructure.

What is AWS Outposts?

For years, AWS was clear on its messaging that customers should *stop spending money on undifferentiated heavy lifting*. This is AWS verbatim, as can be seen in the design principles for the Cost Optimization pillar of the *AWS Well-Architected Framework* (`https://docs.aws.amazon.com/wellarchitected/latest/framework/cost-dp.html`). As it says, racking, stacking, and powering servers fall into this category, with customers advised to explore managed services to focus on business objectives rather than IT infrastructure.

From that statement, it would be reasonable to conclude that AWS would hardly give customers an offering that could resemble the dreaded kind of equipment that needs power, racking, and stacking. The early strides of AWS bringing physical equipment to customers were in the form of the AWS Snow family: AWS Snowball Edge devices and their variants (computing, data transfer, and storage).

It does sport the title of being the first product that could run AWS compute technology on customer premises, being able to deliver compute using specific **Amazon Elastic Compute Cloud** (**EC2**) instance types and the AWS Lambda function, locally powered by AWS IoT Greengrass functions. Despite this fact, it was advertised as a migration device that enabled customers to move large local datasets to and from the cloud, supporting independent local workloads in remote locations.

In addition, Snowball Edge devices can be clustered together and locally grow or shrink storage and compute jobs on demand. AWS Snowball Edge supports a subset of Amazon **Simple Storage Service** (**S3**) APIs for data transfer. Being able to create users and generate AWS **Identity and Access Management** (**IAM**) keys locally, it can run in disconnected environments and has **Graphic Processing Unit** (**GPU**) options.

Launched in 2015, the first generation was called AWS Snowball and did not have compute capabilities, which would appear in 2016 when the product was rebranded as Snowball Edge. Today, AWS Snowball refers to the name of the overall service. The specs are impressive, with 100 GB network options and the ability to cluster up to 400 TB of S3 compatible storage. SBE-C instances are no less impressive, featuring 52 vCPUs and 208 GB of memory.

AWS invested a great deal to make the cloud not only appealing but also accessible. Remove that scary thought of having to change something drastically and radically, that awful sensation of having to rebuild the IT infrastructure on top of a completely different platform. AWS even gave various customers a soft landing and easy path to AWS when they announced (`https://aws.amazon.com/blogs/aws/in-the-works-vmware-cloud-on-aws/`) their joint work with VMware in 2016 to bring its capabilities to the cloud, which debuted in 2017 (`https://aws.amazon.com/blogs/aws/vmware-cloud-on-aws-now-available/`).

With these capabilities and *Edge* appended to the service name, it seemed that moving forward, the path was set with Snowball. It was not without surprise that AWS Outposts was announced in November 2018 during Andy Jassy's keynote at *re:Invent*. On stage, it was shown as a conceptualized model, but one could clearly see it had the shape and form of a server rack.

AWS Outposts debuted on video in 2019 (`https://youtu.be/Q6OgRawyjIQ`), introduced by Anthony Liguori, the VP and distinguished engineer at AWS. By that time, it became clear that a server rack was in the making inside AWS and it was targeting the traditional data center realm. However, it was against the AWS philosophy of asking customers to *stop spending money on traditional infrastructure*. Anyone staring at an AWS Outposts rack could be intrigued.

At re:Invent in 2019, Andy Jassy revealed the use case for Outposts during his keynote. He started by acknowledging some workloads that would have to remain on-premises because even companies who had been strong advocates for cloud adoption had also struggled at times to move certain workloads that proved to be very challenging, and eventually stumbled along their way.

Outposts was characterized as a solution to *run AWS infrastructure on-premises for a truly consistent hybrid experience*. The feature set was enticing: the same hardware that AWS runs on its data centers, seamlessly connecting to all AWS services, with the same APIs, control plane, functionality, and tools as used when operating in the Region. On top of it, it is fully managed by AWS. In the same opportunity, he showcased one specific Outposts variant for VMware, which was a bold move for a cloud company advocating to *stop investing in data centers*.

That was not the only announcement targeting the edge space. At that same event, AWS Local Zones and AWS Wavelength were announced. While these offerings fall beyond the scope of this book, it's worth noting that they weave together to compound an array of capabilities to address the requirements and gaps in the edge space and get a strong foothold in it. So, it suffices to say, AWS Local Zones are built using slightly modified (multi-tenant) AWS Outposts racks.

Now, we have finally set the stage to introduce AWS Outposts. Let us begin with the product landing page (`https://aws.amazon.com/outposts/`). At the time of writing, it is now dubbed *Outposts Family*, due to the introduction of two new form factors at re:Invent in 2021. The 42U Rack version, the first to be launched, is now called an *AWS Outposts rack*. The new 1U and 2U versions are called *AWS Outposts servers*.

Regardless of family type, three outstanding statements that are valid across the family and strongly establish the value proposition of this offering:

- **Fully managed infrastructure**: Operated, monitored, patched, and serviced by AWS

- **Run AWS Services on-premises**: The same infrastructure as used in AWS data centers, built on top of the AWS Nitro System

- **Truly consistent hybrid experience**: The same APIs and tools used in the region, a single pane of management for a seamless experience

Let us cover each in detail.

One of the key aspects of positioning AWS Outposts in customer conversations revolves around explaining how AWS Outposts is different from ordering commodity hardware from traditional hardware vendors. That is exactly where these three statements come into play, highlighting differentiators that cannot be matched by competing offerings.

AWS Outposts is fully managed by AWS. While others may claim their products are also fully managed, AWS takes it to the ultimate level: it is an AWS product end to end. The hardware is AWS, purchase and delivery are managed and conducted by AWS, product requirements are strongly enforced by AWS, and site survey, installation, and servicing are conducted by AWS. No third parties are involved – the customer's point of contact is AWS.

AWS Outposts enables customers to run a subset of AWS services on-premises and allows applications running on Outposts to seamlessly integrate with AWS products in the region. Single-handedly, the first line itself knocks out traditional hardware. For example, you can't run EC2 on it. To amend the case, while applications running on traditional hardware can interact with AWS via API calls, AWS Outposts once again takes it to a whole new level, stretching an AWS Availability Zone in a given Region to the confines of an Outposts rack, allowing workloads to operate as if they lived in the same Region.

Customers are extremely sensitive to consistent processes. The use of multiple tools, multiple management consoles, and various scripting languages is cumbersome and error-prone. When you craft a solution where multiple parts come from multiple vendors and are all assembled, that is what ends up happening.

You will need to use a myriad of tools, interfaces, and scripts to configure and make it work. Long and complex setup processes, multiple vendors involved in troubleshooting errors, and multiple teams conducting various stages of the process lead to inefficiency, inconsistency, security problems, and significant delays in being ready for production.

IT professionals normally try to avoid this pitfall by pursuing a solution provided by a single vendor, even with the risk of the infamous *single vendor lock-in*. However, one hardware provider hardly ever designs and manufactures all the constituent technologies involved, such as the compute, storage, networking, power, cooling, and rack structures. More often than not, the OEM of some of the components is a third-party vendor, if not a third-party brand itself. In the end, these solutions are a *collection of individual parts* with some degree of consistency.

Here is another significant differentiator of AWS Outposts, which is a thoroughbred AWS solution. AWS Outposts employs the same technology used in AWS data centers whose hardware designs and solutions have undergone significant advancements over time and have been battle-tested in production for several years. With this level of integration and control, AWS can explore and tweak the components for highly specialized tasks, as opposed to the more *general-purpose* approach of commodity hardware.

AWS developed a technology called the AWS Nitro System (https://aws.amazon.com/ec2/nitro/), which is a set of custom **application-specific integrated circuits** (**ASICs**) dedicated to handling very specialized functions. AWS Outposts uses the same technology, standing in line to receive any of the latest and greatest advancements AWS can bring into the hardware technology space. Being such a uniform and purpose-built solution, it benefits from a fully automated, zero-touch deployment for maximum frictionless operations.

Now, we are equipped to widely understand the AWS Outposts offering as a stepping stone deployed outside the AWS cloud, with strong network connection requirements to an AWS Region, capable of running a subset of AWS services and capabilities, and conceived and designed by AWS with its own DNA.

AWS Outposts is not a hardware sell, it is not a general-purpose infrastructure to deploy traditional software solutions, and it is not meant to run disconnected from an AWS Region. AWS Outposts is a cloud adoption decision because you are running your workloads not in a cloud-like infrastructure but rather, in a downscaled cloud infrastructure. This is evident because, during the due-diligence phase, an AWS Outposts opportunity can be disqualified by the field teams if the customer workloads are capable of running in an AWS Region. AWS believes in the philosophy that if workloads are capable of running in AWS Region, they should run in an AWS Region.

Basically, AWS is asking what the use cases and business requirements are that prevent certain workloads from operating in the cloud, something that could defy common sense. Does that mean AWS is trying to discourage the customers from taking the Outposts route in favor of bringing them from the edge to the core Region?

Very much the reverse – AWS wants to make sure customers are making informed decisions. It wants them to understand the use cases for Outposts. Fundamentally, they understand they are effectively setting foot in the cloud with Outposts being the enabler to galvanize cloud adoption and the catalyst for companies to upskill their teams to build a cloud operations model and become trained in AWS technologies and services.

At this point, you should be able to identify the edge IT space, the gap between the cloud and the on-premises data center, and also understand the historical challenges associated with operating infrastructure spanning these significantly different domains.

As the initial solutions to address this problem were not good enough, AWS developed Outposts to be the answer to seamlessly bridging these two worlds. Now, it is time to frame AWS Outposts in this edge space to see how it handles the assignment.

Hybrid architecture tenets

Now that we understand AWS and its purpose, it is time to show how it fits in the so-called *hybrid* space. As we outlined in the first section, this concept has been redefined since it was initially coined with a vision originating from the data center to the outside world, where the cloud infrastructure was out there *somewhere*. From this perspective, the edge was anything outside the data center, with the cloud providers being just one of the alternatives for running a compute.

As the advent of the cloud gained traction, solidified, and became the de facto standard as an IT choice for any business to run its workloads and applications, the movement now originates in the cloud reaching out into the world, where the edge nodes are. Any given data center now represents just one of these nodes where compute processing power takes place.

When designing a product to address specific use cases or be fit for certain purposes, one of the strategies is to define the tenets of the architecture. Simply put, the tenets express a belief about what is important or guide us on how to make decisions, which is vital in helping teams to remain focused on what is most important and move quickly as we scale and deliver on our promises. Most good tenets tell you the *how* and not the *what*.

If correctly defined, tenets help to resolve critical decisions by lowering the cognitive load, promoting consistent decision-making over time and across teams, and effectively educating all involved personnel on the thinking and approach to a problem, which, in turn, produces richer feedback. It is a relieving mechanism to gain velocity in delivering a product and a guiding structure to keep the process on track, avoiding derails.

To define some tenets for the hybrid space, enterprises started thinking about the challenges involved in this space and *how* to address them. Any solution has to be a mechanism driving hybrid adoption and must demonstrate how it meets business demands. It needs to have a clear focus on the business problems that it tries to solve and uniquely describe the use cases it is best suited for.

Modern hybrid cloud infrastructure is starting to sediment around some tenets. Here is a list of commonly identified beliefs for a hybrid cloud architecture:

- Inherently secure
- Reliable and available
- Simplicity of use
- Build once and deploy anywhere
- Leverage existing skill sets and tools
- Same pace of innovation as running in the cloud

Here, we can see the power of the tenets in driving decision-making. For example, if we are confronted with a security decision with multiple options on how to address the requirements, the tenet on simplicity of use comes into play to help rule out solutions with a high level of complexity. This can be a tremendous advantage within *technical debates*.

Tenets are fundamental to narrowing down our choices by creating these soft boundaries on *how* to select, in this example, candidate security solutions and approaches. Only then can the debate focus on *what* to do to utilize any selected choices. Without tenets, there could be turmoil within the process of deciding the ways of doing something, where any decision brings the risk of discomfort and disagreement from some unfavored party.

Now that we have our bedrock, let's frame Outposts into the hybrid space to see how it fits:

- **Inherently secure**: AWS says *security is paramount* and that holds for every AWS product. Any AWS product is secure by default, and it does not grant full, unlimited, or public access unless strictly told to do so. AWS Outposts is no different.

Let's showcase features geared towards making AWS Outposts, as any AWS solution, fully furnished with security capabilities.

On the physical structure, AWS Outposts features a built-in tamper detection, which is powered by the AWS Nitro System. The Nitro System is a set of **peripheral component interconnect express (PCIe)** cards purposely built to offload tasks related to security, network, and storage to be executed by dedicated ASICs and spare CPU cycles.

Outposts comes in an enclosed rack with a lockable door and features the **Nitro Security Key (NSK)**, a physical encryption module with destruction available that is equivalent to physical data destruction (data shredding on the hardware).

In the **data protection** realm, AWS Outposts comes with *encryption at rest* enabled by default. Amazon **Elastic Block Store (EBS)** volumes are encrypted using **AWS Key Management Service (AWS KMS)**, where customers store their **customer master keys (CMKs)**. For Outpost servers, the Amazon EC2 instance store is encrypted by default. Moreover, encryption is required. You cannot create unencrypted EBS volumes.

All data traffic is protected by enforcing *encryption in transit*. Any communications between an Outpost and its AWS Region use an encrypted set of VPN connections called a **service link**. User data, management, and application traffic fall under the responsibility of the customer, but AWS strongly advises using an encryption protocol such as **Transport Layer Security (TLS)** to encrypt sensitive data in transit through the local gateway to the local network.

When an Amazon EC2 instance is stopped or terminated, any memory space allocated to it is scrubbed (as in, set to zero) before it can be reallocated to any other instance. The data blocks of storage are also reset.

In the IAM realm, AWS leverages the capabilities of its IAM service. Authentication and authorization are handled by this service. As mentioned, by default, IAM users don't have permissions for AWS Outposts resources and operations. Any permissions must be explicitly granted by attaching policies to the IAM users or groups that require those permissions.

Being a managed service, AWS Outposts operates under the strict AWS global network security procedures that are described in the *Overview of Security Processes* white paper available at (`https://d0.awsstatic.com/whitepapers/Security/AWS_Security_Whitepaper.pdf`). API calls to interact with AWS Outposts are encrypted with TLS and all requests must be signed using an access key ID and a secret access key associated with an IAM principal.

- **Reliable and available**: Reliable can be characterized in short as something *that may be relied on and is fit to be depended on and trustworthy*. Available can be something that is said to be *capable of producing the desired effect*. Using these dictionary definitions as lenses, let's glance at how AWS Outposts takes shape.

The rack comes with all the expected bells and whistles highly available. Redundancy can be identified across power components and networking gear. When it comes to server units, it is a strategy derived from customer availability requirements. AWS recommends the allocation of additional capacity, especially to operate mission-critical applications. One strategy is to order the rack with a built-in capacity to support N+1 instances for each instance family to enable recovery and failover if there is an underlying host issue.

When you create AWS Outposts, it is tied to a single Availability Zone in a given Region. The control plane exists in the Region and is fully managed by AWS. When designing for failure, benefitting from multiple Availability Zones that exist in any AWS Region, the strategy should involve having individual racks associated with distinct Availability Zones.

This approach can be stretched even further. Designs may take into consideration different power grids, but AWS recommends at least dual power sources in a given location. AWS may also consider different physical locations or place the racks apart in different buildings or a Metro area. Certainly, multiple network connections are something to be strictly enforced, optimally using different providers and making sure they operate using different circuits to deliver communication ports at the customer site.

Lastly, you can leverage EC2 placement groups to make sure compute instances are placed onto distinct hosts in a given rack or distinct racks. This capability works in the same manner as in the Region, supporting *cluster*, *partition*, or *spread* placement groups, where a *cluster* strategy requires a single rack and *partition* and *spread* strategies require multiple racks.

Naturally, the application architecture must offer some high availability or fault tolerance feature – otherwise, it will solely rely on server hardware redundancies and mechanical or electronic components in the **Mean time between failures (MTBF)**. Some third-party solutions promote the ability to replicate operating systems and their changes using tracking features such as **Changed Block Tracking (CBT)**, but those must be tested and validated by the customer.

- **The simplicity of use**: The AWS Outposts experience is designed to be frictionless and streamlined as much as possible. Starting with the outstanding status of a *managed service*, customers don't have to touch the hardware, they don't need to configure anything on the rack, they don't have to connect to serial ports or management modules, and they don't have to replace spare parts. AWS is fully responsible for carrying out these tasks. Customers order the rack, and once it is deployed and paired with its peered Region, they begin consuming its resources. Period.

This may appear to be a simple concept but it is huge. If you ever used traditional hardware in your IT department, you certainly know everything that comes associated with it. The selection process, while it can be compared to a fun exercise of choosing food from a menu at the restaurant, frequently may turn into a nightmare of comparing long lists of specifications and features because procurement and purchasing departments always want *apple-to-apple* comparisons.

Okay, let's say you survived the quest to find a cheaper price for commoditized hardware that should, ideally, use the same components and deliver the same performance with a similar set of technical specifications, so much so that they could be considered *interchangeable*. You order it, you get your estimated delivery date, it arrives on site, and now the fun begins.

The task of receiving boxes, unpacking and seating racks, racking individual components, setting up wires to bring power and cables to enable the network, and starting up, configuring, and provisioning the services is all yours unless you buy vendor services for that purpose. Again, let's say it all goes well. Now, it is time to put it to the test, starting by carving out resources and interoperating with other components. All good – it worked. End of story? No, not at all!

Now, it is time to maintain the infrastructure. The significant overhead of patching and upgrading the equipment and managing against a complex *compatibility matrix* across various hardware and software components is risky, potentially disruptive, time-consuming, and often nerve-racking. As time goes on, the cycle continues, only to start all over again when the next *hardware refresh* period kicks in.

None of these is an issue for AWS Outposts, as you are always dealing with AWS. Procurement is tremendously simplified – the business has decided to move to the cloud using AWS technology, but there is nothing to compare or match. Delivery, installation, power up, and startup are operated and supervised by AWS.

All the necessary information, preparation, and site readiness are organized in advance to make sure everything works neatly. Technicians are sent on site to verify whether network configurations are correct and automated processes are working as expected. Service readiness can be achieved in a matter of hours by the time Outposts is brought to life. By the time the bootstrap is finished and the logical *Outpost ID* is up and running, you can jump into AWS Console and begin using your AWS Outposts rack. Don't worry: AWS takes care of maintenance.

- **Build once, deploy anywhere**: We are now using Outposts. How does this solution measure against this tenet? As mentioned before, it is not *AWS-like* infrastructure – it is *AWS-downscaled* infrastructure. Its design, components, technical solutions, and capabilities are the same as those used inside AWS Regions. This is mind-boggling – by ordering AWS Outposts, you are effectively bringing a piece of AWS to your site, for your use, at your disposal.

  That translates into the power of consistency. If you have ever used AWS in the region, it is the same experience with AWS Outposts—same console, same concepts, same configurations. If you provision EC2 in the region, it is the same EC2 provisioned in the rack. If you leverage EC2 service capabilities, they will likely be available to be used in the rack. A service that is available on Outposts has very similar capabilities when compared with the same service operating in a region.

  While, understandably, a service running on Outposts will hardly have all the feature sets available in the region, it is absolutely and truly the same foundation. The only difference is the natural constraint of operating within the limits of the Outposts hardware. Within a region, we have that sense of *virtually unlimited* hardware and the hundreds of AWS services available and connected using gigantic network pipes. Within the confines of an Outposts rack, other restrictions and considerations may apply.

  Physical limitations aside, it is powerful to be able to build applications using AWS technologies and solutions present on Outposts and deploy them seamlessly to any rack on any location. No adaptations, changes, conversions, or adjustments are needed. It runs on a given Outposts SKU and it will run on any similar Outposts, anywhere, anytime. It runs on Outposts, so it will run even more comfortably in the Region.

- **Leverage existing skill sets and tools**: This tenet relates to productivity and the DevOps community. There are challenges associated with the usage of different APIs and tools to build apps for the cloud and on-premises environments, and then the need to re-architect them to work in other environments. The question arises on how to build applications once, run on-premises or in the cloud using the same APIs and tools, and use the wide range of popular services and APIs that you use in the cloud for the applications that run on-premises.

  Enter AWS Outposts. You build on Outposts the same way you build in the region. You use the same AWS Console to view and manage their resources, whether those resources and services are in the AWS cloud or on-premises. You can use the same AWS CLI and SDKs to run and deploy applications and use the same API endpoints to send requests to applications running in the AWS cloud.

  AWS services commonly used by applications running in the cloud are also available when these are deployed on Outposts. Foundational components such as IAM policies and permissions, VPCs, security groups, and access control lists work the same way as in a Region.

API calls will automatically be logged via AWS CloudTrail and tools such as AWS CloudFormation, Amazon CloudWatch, AWS Elastic Beanstalk, and others can be used to run and manage applications running on-premises just the same as they are used for cloud workloads today. If you already use CloudFormation, existing templates will also work with minor tweaks.

Businesses are very sensitive to the impacts of productivity. The time to market is vital, as the pace of innovation is tremendous. Competition is fierce and developers must constantly be launching new features and improving applications. Development cycles have been shortened. Multiple commits are made per day, resulting in multiple daily build cycles, often in the magnitude of tens but stretching to hundreds, even thousands.

Outposts can rightfully make the case for increased development productivity because it is a genuine portion of AWS tech reaching out to the customer premises for their private use and not a venture of multiple suppliers and vendors compounding a solution that is delivered to a customer to act as a bridge between both locations.

- **Same pace of innovation as in the cloud**: Here, we consider the challenges associated with supporting the business. Inefficiency in IT results in poor **Return on Investment (ROI)** for technology investments and slower business growth. As a result, on-premises environments lag behind the cloud when considering the pace of innovation. In a fast-paced, ever-changing environment, this can ultimately appear to be the death knell for a company.

How do we leverage new technology innovations for better and faster deployment of services and deeper business insights? How do we enable innovation in your on-premises environments using services that exist in the cloud?

The answer can only be to have a solution at the edge that evolves in close parity with its parent cloud. AWS Outposts reaches a breakthrough, being the only solution that employs the same foundations as its parent cloud company and being able to evolve in close contact with AWS's latest data center advancements.

- **A truly consistent hybrid experience**: This is the statement coalescing the AWS vision about hybrid environments, which ultimately led to the development of AWS Outposts, and it excels in this aspect. If you look at other cloud providers and their proposed solutions to conquer the edge, look at how many can proclaim to be as seamless, comprehensive, and integrated as Outposts.

Are they designing and manufacturing their hardware infrastructure? Do they use the exact same technologies as used in their data centers? Outposts can rightfully make these claims and they hold on both counts. To date, AWS Outposts is an unmatched product when it comes to fulfilling the promise of the *everywhere cloud*.

Now that we have explored the concept of tenets and how they can be helpful to drive, among other things, the development of a product, let's go ahead and see how AWS Outposts as a solution match the tenets for a hybrid space with the help of some use cases.

Use cases for AWS Outposts

AWS has a method for product development called **working backward**. This is their approach to innovation and the stepping stones to creating a new solution. There is an excellent talk recorded during re:Invent 2020 about this mechanism, available at `https://www.youtube.com/watch?v=aFdpBqmDpzM`.

One step of this mechanism involves asking five questions about the customer, as follows:

- Who is the customer?
- What is the customer's problem or opportunity?
- What is the most important customer benefit?
- How do you know what your customer needs or wants?
- How does the experience look?

The process is composed of several steps to assess the opportunity, propose solutions, validate with stakeholders, and finally, build the product roadmap. Permeating all this process is the concept of a *use case*. Simply put, this consists of exercising a hypothetical scenario to determine how a user interacting with the product can achieve a specific goal.

Use cases are so important because they are the North Star guiding product development. A product must be tailored to address the use case, meaning it will validate the scenario and effectively achieve the pursued goal. A very complex and elaborated product created without a clear purpose can be a display of technical prowess and craftsmanship, and can also carry the risk of not being successful because of the inability to describe *what it is good for*.

For those positioning AWS Outposts as a solution, this is one of the significant challenges. The first callout should undoubtedly show pictures of the products, either the rack or server, with no detailed explanations prior. This will unequivocally trigger in the minds of IT professionals peeking at AWS Outposts for the first time, "*AWS is now selling hardware!*" This is not the case and is a very common pitfall.

This is the first opening to make a statement that AWS is *not* selling hardware. Start by saying the hardware does not belong to the customer – this is not an option. Their minds will switch to thinking it is a hardware rental or leasing contract. This is the time to pull the *fully-managed* service card; AWS takes care of absolutely everything and the customer does not touch the hardware.

This is the part where customers switch to thinking about the legacy *hosting* model, believing that AWS is now supplying **Hardware as a Service (HaaS)**. Certainly, there is a taste of this model in AWS Outposts, but the trump card to be played is simply the statement that you can't just run your platform of choice in it – you only run AWS Services. No commercial hypervisors and no bare metal servers to install on your preferred operating system. It runs the AWS platform.

At this point, you have paved the way to go full throttle into what AWS Outposts aims for at its core: taking the AWS Outposts route effectively means you decided to move to the cloud with AWS. You opened the door for AWS to establish an embassy in your territory to work in close cooperation with you and the crucial reason for this decision should be that you are already looking forward to using AWS services.

If long-term, AWS is not in your equation and you are just looking at AWS Outposts as a potential solution to be used until the next business cycle, where you will re-evaluate cloud providers and look at their similar offerings, trying to make a strong price argument point to justify migrating everything running on Outposts over that other solution, you are potentially treating IT infrastructure as an item in a reverse auction. The cloud provider with the lowest bid wins.

As natural as it may sound, because this is ultimately the *market forces* in action, it may also cast a shadow on these IT departments, implying, from their point of view, that cloud providers are *all the same* and there is no real difference, so they can also be treated as commodities. This statement could not be more naïve; choosing a cloud provider is a decision requiring thorough consideration and an extensive amount of work assessing and evaluating their services and capabilities, combined with a long-term view.

From this aspect, AWS does an excellent job at communicating the value proposition of Outposts and helping customers to make an informed decision. AWS believes in working backward from the *customer's requirements* and wants to be absolutely sure in understanding who its customers are, what the customer's problem is, and what the benefit is for the customer. However, the real deal here is that it goes both ways, and AWS also wants to make sure the customer thoroughly understands what selecting AWS Outposts as their answer for the hybrid challenge means.

Outlining the use cases for Outposts, let's break them down into customer problems and opportunities, following Amazon's customer obsession method, the second question listed earlier.

Customer problems

These are the reasons and forces preventing or invalidating the cloud as an option for running the workload. In this category, we can cite the following:

- Latency sensitive applications
- Local data processing
- Data residency requirements

Let's examine each one in detail.

Latency-sensitive applications

The term *latency* is defined as the time that elapses between a user request and the completion of that request. When a user, application, or system requests information from another system, data packets are sent over the network to a server or system for processing. Once it reaches the destination, it is processed and a response is formed and transmitted back, completing the reply. This process happens many times over, even for a simple operation such as loading a web page on a browser.

There might be several network components involved in order to complete this process and each one adds a tiny delay while forwarding the data packet. Depending on the number of simultaneous transmissions and user requests, the traffic mounts up to a point that these delays become perceptive to the user in the form of wait times. This effect is even worse when the data packets need to traverse long geographical distances.

For the end user requesting information from a website, this translates into a long wait time until the web page finally loads. However, some applications simply rely on low-latency networks to work predictably and smoothly – therefore, this characteristic becomes a requirement. Some applications may require ultra-low latency (measured in nanoseconds, while low latency is measured in milliseconds). Other factors to be taken into consideration are latency jitter (the variation in latency) and network congestion.

Good examples of applications and use cases that require low latency can be found across various industries: life sciences and healthcare, manufacturing automation, and media and entertainment. Use cases encompass content creation, real-time gaming, financial trading platforms, electronic design automation, and machine learning inference at the edge. Let's cite a few:

- **Healthcare**: Surgical devices, **Computerized Tomography** (**CT**) scanners, and **Linear Accelerators** (**LINACs**)

- **Life sciences**: Molecular modeling applications such as GROMACS (`https://www.gromacs.org`), and 3D analysis software for life sciences and biomedical data

- **Manufacturing**: Medical device manufacturing, pharmaceutical and **over-the-counter** (**OTC**) manufacturing, integrations with IoT, a digital twin strategy (`https://aws.amazon.com/iot-twinmaker/faqs/`), **Supervisory Control and Data Acquisition** (**SCADA**), **Distributed Control Systems** (**DCSs**), **Manufacturing Execution Systems** (**MESs**), and engineering workstations

- **Media and entertainment**: Content creation and media distribution (streaming)

- **Financial services**: Next-generation trading and exchange platforms

Local data processing

Some use cases may end up generating large datasets that need to be processed locally. Because of their size, the cost to migrate them to the cloud may be unfeasible because the back-and-forth of pre- and post-processing data between the cloud and the site may end up generating significant egress charges and can also lead to packet loss, resulting in data integrity problems.

Moreover, the time it would take may be unrealistic for the use case and effectively will defeat its purpose. Additionally, customer requirements may dictate processing data on-premises and the ability to easily move data to the cloud for long-term archiving purposes or workloads that may need to be available during a network outage.

The same types of industries mentioned before have use cases with this requirement:

- **Healthcare**: Remote surgery robots, computer vision (for medical image analysis), **Picture Archiving and Communication Systems** (**PACS**), **Vendor Neutral Archiving** (**VNA**) solutions, and taking emergency actions on patients carrying wearable devices capable of making decisions using inference at the edge
- **Life sciences**: Cryo-electron microscopes, genomic sequencers, molecular modeling with 3D visualization (requires GPUs), and **Research and Development** (**R and D**)
- **Manufacturing**: Smart manufacturing (`https://aws.amazon.com/manufacturing/smart-factory/`), site optimization, and predictive maintenance

Data residency requirements

Here, let's briefly examine some of the terminology involved as well. **Data residency** is the requirement that all customer content must be processed and stored in an IT system that remains within a specific locality's borders. **Data sovereignty** is the control and governance over who can and cannot have legal access to data, its location, and its usage.

Various forces are driving this requirement and they are present in many organizations across the public and private sectors. Data residency normally comes from the following:

- The obligation to meet legal and regulatory demands, including data locality laws. This requirement can affect, for example, financial services, healthcare, oil and gas, and other highly regulated industries having to store all user and transaction data within the country's borders, or public entities may be subject to a requirement that data produced by local and national government needs to be stored and processed in that country.
- The organization's business and operating model, where the majority of activities take place within a certain country's geography. In this scenario, the company falls under the financial rules of a national entity, which may require storing or processing some or all of its data within that nation state.

- There may be contractual requirements to store data in a particular country as well. Businesses may have to agree to keep the data of specific customers in a given jurisdiction to meet the data residency requirements of those clients.

- Lastly, it could be mandated for business or public sector entities that certain data must be stored or processed in a specified location due to corporate policy. This mandate could be partially or fully derived from one of the previous drivers.

As some use cases of storing sensitive data on AWS Outposts, we can cite patient records, medical device **intellectual property (IP)**, – as in, copyrights, trademarks, and patents – government records, genomic data, and proprietary manufacturing info.

Customer opportunities

These are potential ways to use a product that can propel, expedite, or catalyze the cloud as an option to run a workload. They are potential uses of the product that can help businesses to strengthen their arguments for building a hybrid cloud by adding more strategic use cases. Let us look at some of them:

- Application migration to the cloud

- Application modernization

- Data center extension

- Edge computing

Let's examine each one in detail.

Application migration to the cloud

This may not be immediately perceived as a potential use case, but it turns out to be a powerful one. Large migrations from on-premises data centers to AWS may involve a myriad of applications and can take several years. The risk involved is tremendous if the environments are significantly different, not to mention the operational burden to use multiple management tools, APIs, and interfaces.

AWS Outposts can significantly mitigate, if not eliminate, this problem. Because it is a portion of AWS, it provides a consistent operational environment across the hybrid cloud while migrating applications to the cloud, ensuring business continuity. Your workloads will not need tweaks and adjustments – if they run on Outposts, they will run in the Region just as well. The only point of attention is the strength and sensitivity of its ties with on-premises services.

This is achieved by employing a strategy called two-step migration. Instead of having to migrate applications and critical dependencies all at once, AWS Outposts offers a safe haven where you can begin migrating in steps to Outposts while keeping close contact with the on-premises applications. This enables customers to slowly move all individual components into Outposts and when they are all together, you can easily move to the region.

Still, in the migration realm, AWS offers a tool to expedite migrations called CloudEndure (`https://www.cloudendure.com/`) While it is also a disaster recovery tool, CloudEndure allows all migration paths: from on-premises servers (whether physical or virtual) to AWS Outposts, from AWS Regions to Outposts, from other clouds to Outposts, and even from Outposts to Outposts. Recently, AWS launched a new service for migrations called AWS Application Migration Service (`https://aws.amazon.com/application-migration-service/`), which is the next generation of CloudEndure migration that will remain available until the end of 2022.

Moreover, there is an Outposts flavor that runs VMware Cloud on AWS. VMware customers can easily and seamlessly interoperate and migrate their existing VMware vSphere workloads while benefiting from leveraging their investments on the VMware platform.

Application modernization

Modernizing while you are still on-premises may be the best approach for some workloads that are tightly coupled to the existing infrastructure. There are many opportunities in this area, such as moving legacy monolithic workloads to containers, modernizing mainframe applications, and enabling CI/CD and a DevSecOps approach. AWS Outposts offers the ability to run Amazon ECS or Amazon EKS on-premises to power this transformation.

Modernization with AWS Outposts can be the first step towards the bold objective of *re-invention*. At this stage, customers can use AWS Lambda at their disposal and explore serverless containers with AWS Fargate for both Amazon ECS and Amazon EKS.

Mainframe modernization stands out from the crowd because of the powerful driving forces behind it. Cost savings is the first and the most obvious, the obsolescence of the platform and the business risk it represents are also there, while the ever-growing shortage of skilled professionals to support this legacy is well-known and the reason for some amusing stories.

One particular driving force that normally falls off the radar is the constraints mainframes impose on businesses preventing them from using modern technologies. Keeping the business locked into the limitations imposed by mainframes can be the poison holding companies off from unlocking their market potential.

Data center extension

In this realm, the infrastructure of your cloud provider is treated as an extension of your on-premises infrastructure. This gives you the ability to support applications that need to run at your data center. There are four broad use-cases:

- **Cloud bursting**: In this application deployment model, the workload runs primarily in on-premises infrastructure. If the demand for capacity increases, you branch out and AWS resources are utilized. There are two main reasons triggering the need for cloud bursting:

- **Bursting for compute resources**: You consume a burst compute capacity on AWS through Amazon EC2 and managed the container services of Amazon ECS, Amazon EKS, and AWS Fargate.

- **Bursting for storage**: In this case, you can integrate your applications with Amazon S3 APIs and leverage AWS Storage Gateway. This offering enables on-premises workloads to use AWS cloud storage, which is exposed to on-premises systems such as network file shares (File Gateway) for file storage or iSCSI targets (Tape Gateway and Volume Gateway) for block storage.

- **Backup and disaster recovery**: Customers can leverage the power of object storage with Amazon S3 and use data bridging strategies presented by AWS Storage Gateway, back up their applications with AWS Backup, and move or synchronize data between sites and AWS with AWS DataSync. For disaster recovery strategies based on file data hosted premises that need to be transferred to the AWS cloud, you can leverage AWS Transfer for **Secure File Transfer Protocol (SFTP)**.

- **Distributed data processing**: Certain applications can be deployed with functionality split between on-premises data centers and the AWS cloud. In this scenario, we normally expect the low-latency or local data processing components to stay close to the local network on-premises and other components delivering additional functionality to reside on AWS. In the cloud portion, you can benefit from a myriad of services such as massive asynchronous data processing, analytics, compliance, long-term archiving, and machine learning-based inference. These capabilities are powered by services such as AWS Storage Gateway, AWS Backup, AWS DataSync, AWS Transfer Family, Amazon Kinesis Data Firehose, and **Amazon Managed Streaming for Apache Kafka (Amazon MSK)**, which act as enablers to use the imported data as the source for analytics, machine learning, serverless, and containers.

- **Geographic expansion**: AWS is constantly expanding and evaluating the feasibility to deploy new Regions across the globe, but it's unrealistic to expect regions to be deployed to the tune of thousands or even hundreds of locations. It may need to deploy an application in a place where you are still unable to leverage an AWS Region. Eventually, there might be reasons why workloads need to stay close to your end users, such as low latency, data sovereignty, local data processing, or compliance. Traditional approaches such as deploying your own physical infrastructure can become challenging, costly, or constrained by legal requirements and local laws, but AWS Outposts can be instrumental in fulfilling this use case if it is available in that geography. This information is easily accessible on the product FAQ page (`https://aws.amazon.com/outposts/rack/faqs/`).

Edge computing

Certain environments such as factories, mines, ships, and windmills may have edge computing needs. Outposts addresses this use case with its smallest form factor, Outposts servers – these scenarios are unlikely to be addressed with the Outposts rack. However, the connection requirement to one parent region is still there. When the requirements specifically involve harsh conditions, disconnected operation, or air-gapped environments, customers can use AWS Snowball Edge computing. These ruggedized devices are capable of operating while fully disconnected and use Amazon EC2 compute resources to perform analytics, machine learning, and run traditional IT workloads at the edge. Data can be preprocessed locally and further transferred to the AWS cloud for subsequent advanced analysis and durable retention.

Another edge computing offering is AWS IoT Greengrass, which you can run on Outposts servers. Edge applications generate data that may need to be consumed locally to identify events and trigger a near real-time response from onsite equipment and devices. With AWS IoT Greengrass, you can deploy Lambda functions to core devices using resources such as cameras, serial ports, or GPUs. Applications on these devices will be able to quickly retrieve and process local data while remaining operational, withstanding fluctuations in connectivity to the cloud. You can optimize the cost of running apps deployed at the edge using AWS IoT Greengrass to analyze locally before forwarding to the cloud.

Closing this use cases section, it is worth highlighting the uniqueness of AWS Outposts. This is a product designed with one clear statement in mind: it must be, as much as possible, a portion of an AWS data center stretching out to a customer facility. This paradigm drove product development and will drive product evolution. Anyone using AWS Outposts expects nothing other than AWS technology and this expertise being applied to the product so it can become increasingly valuable.

If we look at the pace of the innovation of AWS, how innovative and visionary their teams are, and how resolute AWS is in advancing with speed and strength without being careless or resting on its laurels, we can safely say that the best is yet to come.

Summary

In this chapter, you learned about the rise of the hybrid IT space and how it transitioned from an approach expanding out from the data center toward the cloud to a movement where the cloud is an expanding stronghold with precedence over data centers.

Next, you learned about AWS Outposts and how it was born to seamlessly bridge the AWS cloud with customer data centers. Then, you were introduced to the concept of tenets, and we defined the tenets for a hybrid cloud solution.

We contrasted AWS Outposts against these tenets to assess how it compares and highlighted use cases for this product. Now, it's time to take a peek at the real thing and take a tour of an AWS Outposts rack. With the knowledge you gained in this chapter, you now have the foundation to understand why AWS Outposts was designed and engineered the way it was, which you will see in greater detail in the next chapter.

2
AWS Outposts Anatomy

At first glance, AWS Outposts looks just like traditional commodity hardware. There is a rack made up of individual components that have the shape and form of servers and networking gear. What could AWS have done to make it so different from the existing offerings from traditional vendors?

As we progress through this chapter, you will get a glimpse of the process to select a suitable datacenter platform. In particular, if you are someone with experience of deploying commercial data center solutions and designing the underlying hardware necessary to support their functionality, may sure to pay attention to how this process unfolds and what is relevant for selecting the appropriate platform.

This chapter will demonstrate what distinguishes AWS Outposts from other offerings, from the hardware it is built with to its capabilities. You will obtain knowledge of the following:

- Structural elements – rack and power
- Communications elements – networking
- Capabilities – services and features
- Connectivity elements – cables and connectors

Structural elements – rack and power

Now that we have a deep understanding of the *edge* space and how AWS Outposts addresses use cases in this realm, it is finally time to take a look at Outposts itself and break down each part. Before we jump into that, it is important to underline the statement we made in the previous chapter: *AWS Outposts is not commodity hardware; it is a downscaled extension of the AWS cloud infrastructure for your data center.*

This is important because a different mindset is required to get the most from this chapter. Architects designing solutions using commodity hardware often analyze certain aspects of servers, storage, and networking that are irrelevant when designing a system with AWS Outposts. I definitely want to discourage your thought process from leaning toward the following:

- Processor models and features, number of cores, L2 and L3 caches, stepping, and families

- Networking chipsets and features

- Drive types, sizes, interfaces, and controller caches

- Redundant fonts, RAID levels, SDCC/DDDC memory, dual-port NICs in segregated slots, dual controllers, SLAs, and time-to-repair

Thinking about these points will certainly derail this chapter. None of this is necessary to be considered in this chapter, nor it is relevant to our conversation. We discuss requirements, capabilities, and services.

One piece of information can illustrate and shed light on how AWS Outposts was conceived and engineered to closely resemble the requirements of large data center environments: this relates to the definition of the **Field Replacement Unit (FRU)**.

Conceptually, this is an item that can be identified as non-operational by performing a set of standard troubleshooting procedures, and can be replaced in the field by a technician or authorized service party. The defective part is sent back for repair or disposal and directly replaced with a new one.

This matters a lot because the breakdown of a complex system into smaller serviceable parts directly influences its cost and availability. Modular construction tends to prefer grouping several individual components into bigger units to simplify the parts' procurement and the knowledge required by technical support, and facilitate reader-friendly documentation.

With AWS Outposts, the defined FRUs are the entire server or the **Outposts Networking Device (OND)**. As a comparison, it's the difference between having to supply and train staff to replace dozens of parts in commodity hardware versus replacing an entire single server. You don't have to worry whether your server hardware will be happy with the new item or whether it will be compatible with all the other components. You will simply receive a whole brand-new server.

Of course, you are now thinking you should architect for the failure of an entire host and somehow have the spare capacity to cope with this possibility. This is absolutely correct because it falls completely in line with AWS messaging: *Everything fails all the time.*

We must embrace failure, we must count on it. As remote as this possibility may look at first, good architecture practices command you must first define your requirements and design your architecture to meet your requirements. When we are in the region, we talk about spreading your resources into different **Availability Zones (AZs)** or even different Regions. With AWS Outposts, we talk about having spare servers, multiple racks, and multiple sites.

Rack

The rack is the foundational structural piece and it is simply described as an *industry-standard 42U rack form factor*, but this does not do it justice. Enough talk, let's look at the visuals of the product.

We will use the invaluable help of the resources and images that can be found at `https://m.kaon.com/c/az`. All the images and views come from this site and are the property of AWS under the AWS Site Terms (`https://aws.amazon.com/terms/?ncl=f_pr`). In this section, let's explore the rack and its power supply. Naturally, this is only present in the Outposts rack unit, and there is no rack in the Outposts server unit.

Figure 2.1 – Outposts rack

AWS Outposts is delivered as a fully-assembled industry-standard 42U rack, with all the components installed, all elements interconnected, and ready-to-go external connections. The rack has a lockable door with built-in tamper detection. Supplied with casters, it is ready to be rolled into the final position where it will be seated using stands. While it does not look like much, there are intrinsic messages to be read between the lines here.

For starters, AWS does not ship boxes of components to be assembled at the site – the whole rack is shipped preassembled. This relates to AWS product engineering and the market messaging around it, clearly demonstrating how different this offering is from commodity hardware. There's no chance of missing or damaged parts, no chance of malfunction due to sloppy assembly processes, and no chance of component incompatibility. It just works, right away.

Second, it sets the bar for customer experience. It has to be, as much as possible, flawless but the customer needs to be ready too. A large transport will arrive and unload a full rack (possibly weighing almost 1 ton) that has to be either positioned in place or stored somewhere, and this is no small task.

You will need a loading dock, enough clearance and weight support along the way, and access to facilities. This alone has significant implications: not all facilities can provide these requirements and there is no option to disassemble the rack. If these conditions are not met, the rack will not be delivered.

Lastly, it sends a strong message about AWS Outposts as a managed service. It is not your equipment, it is not meant to be serviced by you in any way, and you cannot freely install your own components in that rack. AWS Outposts is infrastructure provided by AWS to your facility, dedicated to you, to leverage the AWS services that can be run on that hardware.

Now let's examine the powerhouse. Let's go back in time a little to the AWS re:Invent 2020 keynote delivered by *Peter DeSantis, Senior Vice President of Utility Computing at AWS*, where in the first section there was an extensive talk about reinventing the UPS and AWS technologies around effective and intelligent use of power. Look no further: you can see this technology by looking at AWS Outposts rack.

This talk is available at `https://youtu.be/AaYNwOh9OPg` and at 13:42, you can observe the following still:

Figure 2.2 – AWS re:Invent 2020 - Infrastructure Keynote with Peter DeSantis

Let's take a peek at what you get with your AWS Outposts rack unit:

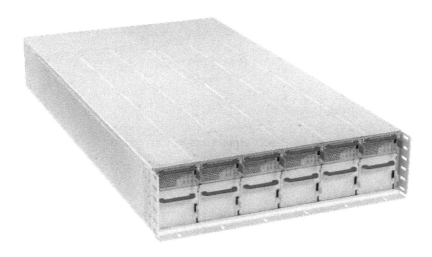

Figure 2.3 – AWS Outposts rack power shelf

As you can see, it is exactly the same piece as in *Figure 2.2*. AWS indeed sticks to its promise that Outposts rack is the same technology used in their data centers. This is also reflected in the AWS Outposts rack hardware specs, found at this URL: `https://aws.amazon.com/outposts/rack/hardware-specs/`.

Figure 2.4 – AWS Outposts rack hardware specs

Power

Power is to AWS Outposts what blood is to the human body. We need a stable flow of power in enough quantity to bring AWS Outposts to life, and it's important to stress that power requirements are stringent.

If your data center can deliver, for the sake of an example, 7.5 kW maximum and you order an AWS Outposts configuration that draws 10 kW power, you will have to split your order into two racks. That may seem like a disadvantage at first, but in *Chapter 5, Security Aspects in Outposts*, we will talk about resilience and how this may come in handy. You can associate these racks with different Availability Zones.

Following are the available configurations to date for the power shelf inputs:

	Redundant, single phase	Redundant, three phase	Single phase	Three phase
5 kVA	2 x L6-30P or IEC309, 1 drop to S1 and 1 drop to S2		1 x L6-30P or IEC309, 1 drop to S1	
10 kVA	4 x L6-30P or IEC309, 2 drop to S1 and 2 drop to S2	2 x AH530P7W or AH532P6W, 1 drop to S1 and 1 drop to S2	2 x L6-30P or IEC309, 2 drop to S1	1 x AH530P7W or AH532P6W, 1 drop to S2
15 kVA	6 x L6-30P or IEC309, 3 drop to S1 and 3 drop to S2		3 x L6-30P or IEC309, 3 drop to S1	

AC line voltage	Single phase 208 to 277 VAC (50 or 60 Hz) or three phase 346 to 480 VAC (50 to 60 Hz)
Power consumption	5 kVA (4kW), 10 kVA (9 kW), or 15 kVA (13 kW)
AC protection (upstream power breakers)	30 or 32 A
AC inlet type (receptable)	Single phase L6-30P (30A) or IEC309 P+N+E, 6 hour (32 A), three phase AH530P7W 3P+N+E, 7 hour (30A), or three phase AH532P6W 3P+N+E 6 hour (32 A)
Whip length	10.25 ft (3 m)
Whip - Rack cabling input	From above or below

Figure 2.5 – Power shelf inputs (S1 and S2) valid configurations

Currently, AWS Outposts does not support DC power and when working with the **single phase** option, you should consider it can use up to six whips in total for 15 kW (basically two whips per 5 kW). If your power receptacles are not supported, you will have to work with AWS personnel to devise a possible solution.

The AWS Outposts power shelf does not come with **Battery Backup Units** (**BBUs**) because it will be ideally deployed in a data center with some form of UPS system. It has a total of six **power supply units** (**PSUs**) but they work in steps of 5 kW (or 5 kVA), meaning depending on the power draw of your rack, not all PSUs might be required. Once seated, the power shelf takes the center position on the rack.

Because the power supplies work in steps of 5 kW, if you select a rack configuration with intermediary power consumption it will be rounded up to the next plateau (for example, if your configuration draws 6 kW, you must provide capacity for 10 kW).

The AWS Outposts server option has far simpler power requirements, as described on the product requirements page (`https://docs.aws.amazon.com/outposts/latest/userguide/outposts-requirements.html`): *Servers are rated up to 1600 W 90-264 VaC 47/63 Hz AC power*. Outposts server is shipped with power cables, but the power supply receptacle is suitable for a rack power bar; to plug it into a wall outlet, an adapter is needed. It has redundant power supplies in the form of two independent power units that have to be connected to different power plugs wired to distinct power circuits.

These units work in an active-active arrangement, and in case of failure of one of the units the other unit is capable of keeping the server running without interruption. The server is capable of working with just one of the power units plugged in. A single power outlet is not recommended because a power outage could cause the server to fail with unpredictable consequences.

Now let's take a look at the rack bus bar at the back of the rack.

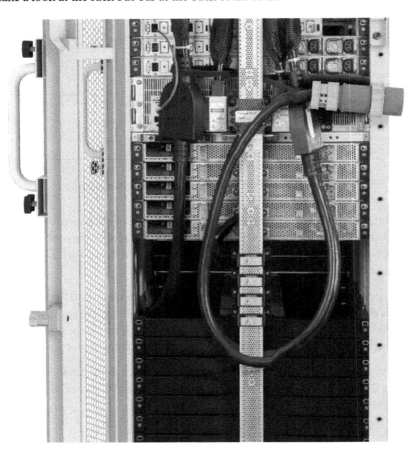

Figure 2.6 – AWS Outposts rack bus bar

The bus bar runs vertically along the middle of the backside of the rack and directly links with the power shelf. There is a significant amount of power running through this bar and one must be very careful working around it. Each server connects directly to this bar, eliminating the need for power cables.

Let's take it back a bit to reflect on how different this is from commodity hardware. Purpose-built technology, cutting-edge engineering solutions, sustainable and future-proof design, and end-to-end control over the technology, enabling the solution to be fully owned and supported by one vendor: and we've only seen the skeleton and the blood that runs through Outposts' veins, so hopefully, you are excited for what's next.

In this section, we examined two fundamental aspects of the AWS Outposts *body*: The *skeleton* (rack) and the *blood in its veins* (power). Now that we are able to bring our AWS Outposts rack to life, let's look at how it communicates with the external world. In the next section, we will talk about networking in depth.

Communications elements – networking

It is time to understand the communication pathways. This is certainly one of the most intriguing components of Outposts. If you understand how AWS networking works, with Regions and components including VPC, Security Groups, and Network ACLs, then you know that AWS networking is a stunning piece of communications engineering that unsurprisingly extends to AWS Outposts.

Let's look at one physical networking switch:

Figure 2.7 – An Outposts Networking Device (OND)

AWS will run its specific features and leverage Juniper technologies for the switch to deliver the underlying infrastructure. For AWS Outposts, server connectivity is far simpler, as it does not come with any switch; this component is the responsibility of the customer. You have the option to go with two 1-GBps RJ-45 ports or one QSFP+ port with a breakout cable with four SFP+ connectors – two are operational and the other two are disabled.

AWS Outposts seamlessly extends VPCs for its communication with its parent Region via a *Service Link*. As mentioned before, a *Service Link* comprises a set of encrypted tunnels connecting AWS Outposts to its *anchor*, which pairs Outposts devices with services living in the AWS Region, and, therefore, AWS Outposts is not meant to run disconnected for extended periods of time.

Let's talk about this umbilical pairing of AWS Outposts with a parent region. This is crucial to design solutions using Outposts. The following hierarchical series of statements will clarify how Outposts is tied to AWS Regions:

- AWS Outposts can be paired with just *one* Region and with just *one* Availability Zone in that Region. It can be any Region where the service is offered (check this URL: `https://aws.amazon.com/about-aws/global-infrastructure/Regional-product-services/`).

- *Multiple* VPCs can be extended to AWS Outposts; however, you can *only* create subnets on AWS Outposts in the Availability Zone that your unit is associated with.

- You can create *internal VPCs* that exist only inside AWS Outposts and as expected you can create subnets that only exist inside AWS Outposts.

- You can leverage the default route (or quad-zero route) `0.0.0.0/0` to route traffic to your local network rather than routing traffic to the Region, which is the default configuration.

- Traffic from the Region traversing AWS Outposts to reach local networks is explicitly forbidden.

- There are no charges for *data out* (*egress*) from AWS Outposts to the Region, but *data in* (*ingress*), which is traffic from the Region directed to Outposts, is charged.

Common questions around the Service Link include the following:

- What is the minimum required bandwidth?

  - **Outposts rack**: Connectivity of at least 500 Mbps (1 GBps is better) with sub-150 ms latency. Redundant connections are strongly advised.

  - **Outposts servers**: Connectivity of at least 20 Mbps.

- What type of traffic flows between AWS Outposts and its parent AWS Region?

  - **Management traffic**: Telemetry, state, and health data about various components, maintenance tasks originating from AWS (such as software and firmware updates, system commands, and traffic destined to Amazon CloudWatch and AWS CloudTrail), API calls made by a customer to configure services on AWS Outposts, AWS IAM forwarded actions, and DNS replies.

  - **Customer traffic**: Any network traffic or interaction between resources living in the Region and resources living in AWS Outposts. It is the customer's responsibility to manage network communications between instances and services running on their Outposts and those running in the AWS Regions.

- Is AWS **Direct Connect** (**DX**) required? – No, you can use any internet connection available from your service provider that meets the requirements.

- If I have Direct Connect, what types of connectivity are supported? – Both public (public VIF) and private (private VIF) exclusively.

- What is the minimum MTU required across an on-premises network? – 1,500 bytes.

- Available routing protocols: BGP (static routes can be used).

- Required firewall port: 443.

- Can I block inbound connections? – Yes, AWS Outposts initiates all VPN connections. If you use a stateful firewall, then you do not need to allow inbound connections; they are allowed by default. Stateless firewalls require configuration to allow returning traffic (where the AWS Region replies to AWS Outposts-initiated traffic).

- What happens when the Service Link connection is down?

  - Running EC2 instances and their EBS volumes will continue to function, and they can be reached via the local network.

  - Running services (EKS worker nodes, RDS Instances) will continue to run; however, they will degrade over time. Mutable API calls (e.g. Run/Start/Stop/Terminate) will not work because they depend on the Region.

  - IAM authentication will not work, and existing tokens and authorizations will expire over time.

  - You will be unable to access your objects stored in S3 on Outposts as it depends on IAM for authentication and authorization.

  - DNS queries forwarded to the Route 53 resolver in the Region will not work. You should strongly consider having local DNS resolution if you expect to have short periods of disconnected/limited bandwidth during your operations.

  - Metrics and logs will be stored locally for a few hours and pushed to the AWS Region once connectivity resumes.

More information about connectivity to the Region can be found here: `https://docs.aws.amazon.com/outposts/latest/userguide/how-outposts-works.html#Region-connectivity`.

In short, AWS Outposts is similar to having a small portion of an AZ infrastructure deployed on premises with all the capacity dedicated to you. You can run a subset of services that exist in the Region and leverage the same APIs and tools that you use in the Region. You manage Outposts directly on the AWS Management Console and it is a truly seamless experience.

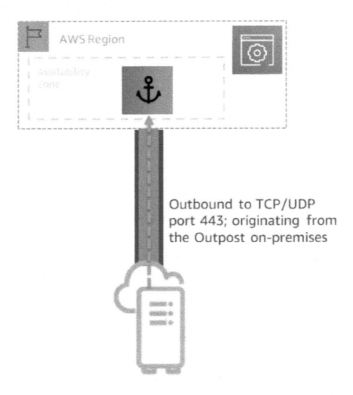

Figure 2.8 – AWS Outposts Service Link connection to an AWS Region

At this point, we understand that AWS Outposts needs to reliably connect to an AWS Region, and your experience in terms of launching AWS services, sending API calls, and interacting with services in the Region will be strongly dependent on the available Service Link bandwidth. Operating AWS Outposts with less than 500 Mbps or with high latency could give you an undesirable user experience. For example, launching a big AMI inside Outposts may take a long time because this AMI will be downloaded from the AWS Region to AWS Outposts via a Service Link.

Networking resources

Let's identify the networking resources using visuals. On the left side, we depict a representation of an AWS Region. You can identify the typical networking services and capabilities present in the Region: **Virtual Private Cloud (VPC)**, **Internet Gateway (IGW)**, **Transit Gateway (TGW)**, **VPC Peering (PCX)**, **VPC Endpoints (VPCE)**, **Virtual Private Gateway (VGW)**, and **subnets** (public and private).

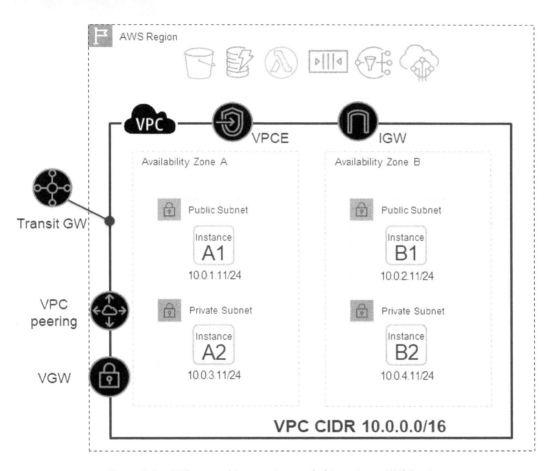

Figure 2.9 – AWS networking services and objects in an AWS Region

The Service Link connects AWS Outposts with a single AZ in a given parent Region of choice and allows the VPC to be extended to AWS Outposts as shown in the following diagram.

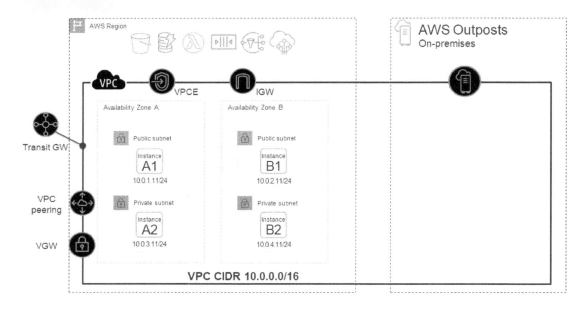

Figure 2.10 – VPC in the Region extended to AWS Outposts

Inside this VPC, we can create subnets associated with the AZ in which the AWS Outposts unit is anchored, and these subnets will exist in the Region. However, if we want to create a subnet that will be associated with AWS Outposts, we need to use the AWS Outposts console (under **Actions | Create subnet**), as shown in the following screenshot.

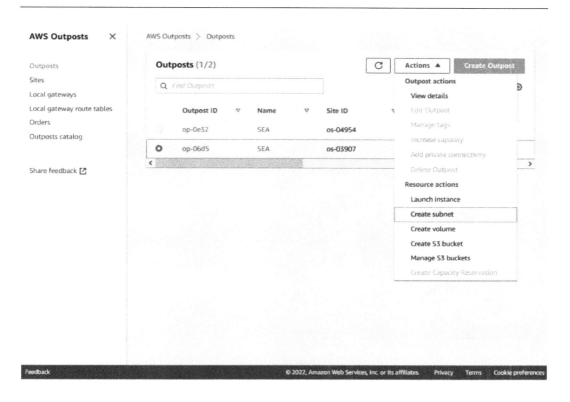

Figure 2.11 – AWS Outposts console "Create subnet" action

We can achieve this using the AWS CLI, but note that an extra parameter (`--outposts-arn`, described in the API reference at `https://docs.aws.amazon.com/cli/latest/reference/ec2/create-subnet.html`) needs to be provided, indicating that the subnet must be created inside AWS Outposts.

Once these subnets are operational, they will appear in the list of the subnets available to launch EC2 instances inside the VPC. For example, suppose the Outposts subnet is selected. In that case, the AWS control plane will *understand* these instances must be launched inside AWS Outposts rack, and it will consume Outposts rack's resources (compute, memory, and storage). Otherwise, the instances will be launched in the Region and consume resources in the AZ corresponding to the selected subnet as usual.

However, AWS Outposts lies io the edge primarily because it needs to interact with services and components that live there; therefore, it needs to connect to the customer's local network, potentially with extremely low latency. To meet this requirement, a new logical construct was needed because the VPC did not have a specific gateway for this network portion. To address that, the **local gateway** (**LGW**) was developed for AWS Outposts, and the **Local Network Interface** (**LNI**) was developed for AWS Outposts servers.

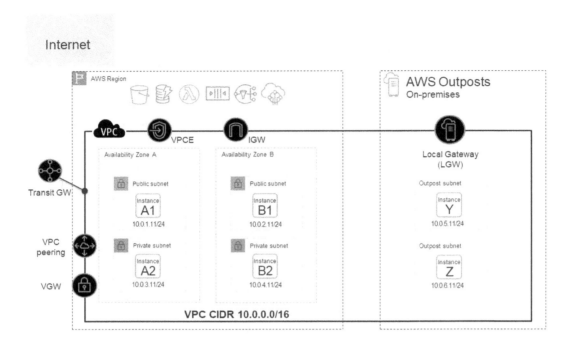

Figure 2.12 – Subnets inside AWS Outposts indicate EC2 instances must be launched on that location

Just as we have route tables associated with subnets in the Region, we have these same objects associated with the subnets that live inside AWS Outposts. The same concepts used in the AWS Region apply to AWS Outposts; we can have routes associated with logical objects in the VPC.

The LGW has two primary roles: first, it provides a local gateway route table (*lgw-rtb-<id>*) that can be associated with VPCs so they can communicate with your on-premises data center. Second, it provides a target to be used in your VPC route tables (*lgw-<id>*) so network traffic can be forwarded to local networks.

Communication between instances running inside Outposts is enabled by default via the *local* target (the *intrinsic* router connecting all subnets created inside the VPC).

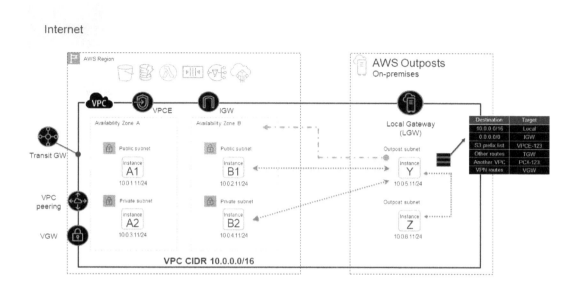

Figure 2.13 – Route tables can be associated with subnets that live inside AWS Outposts

Finally, to enable processes running inside AWS Outposts to communicate with resources that are reachable via routes forwarding packets to local networks, we need to configure these network segments in the route table associated with the local gateway.

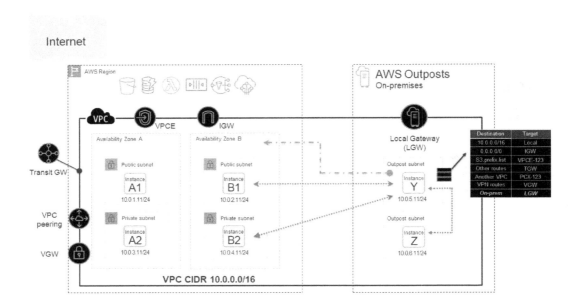

Figure 2.14 – Routes pointing to local network segments must be associated with the LGW

One more step is needed before we have established communication. The LGW is a network process that performs **Network Translation (NAT)**, meaning the EC2 instance running on Outposts has two logical IP addresses: one that belongs to the subnet associated with the Outposts ID and another one that is part of the customer's local network.

This IP address is called a **customer-owned IP address** (**CoIP**) and it behaves pretty much like an *elastic IP address*: You have an IP address pool (referred to as the customer-owned pool) and you associate one elastic IP address from this pool with your EC2 instance. It can be assigned automatically (using the **EC2 Console** under the **Elastic IPs** menu) or you can specify it manually (this action can only be performed using the AWS CLI).

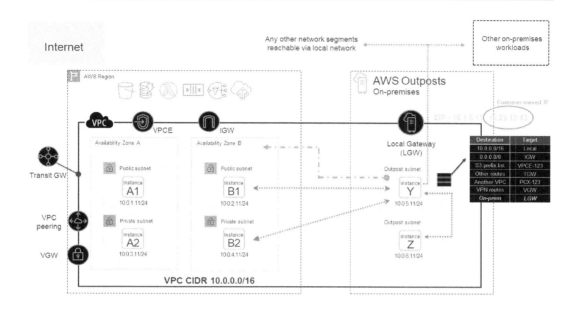

Figure 2.15 – Customer-owned IP performs NAT translation with IPs attached to the VPC subnet

The CoIP address range must be identified by the customer during the ordering process and, obviously, must be a CIDR range that exists inside the local network.

> **Note**
>
> For routing to function properly, all these ranges must neither overlap nor conflict.

This YouTube video (https://www.youtube.com/watch?v=6rSjLCFOz1c) shows this process in detail using the AWS Management Console. You can find more information about the elements described in this section at https://docs.aws.amazon.com/outposts/latest/userguide/how-racks-work.html.

In AWS Outposts servers, there is no translation or logical segmentation. The LNI operates in Layer 2 and gets the IP address from the customer network. Segmentation is physical; we either have two 1-GB ports (where port 1 is the LNI and port 2 is the Service Link) or the QSFP+ breakout cable where the SFP+ connector that can act as the LNI is labeled **1** or **A**, and the SFP+ connector for the Service Link is labeled **2** or **B**.

The following is a depiction of the networking elements for Outposts servers. You can find more detailed information at `https://docs.aws.amazon.com/outposts/latest/userguide/how-servers-work.html`.

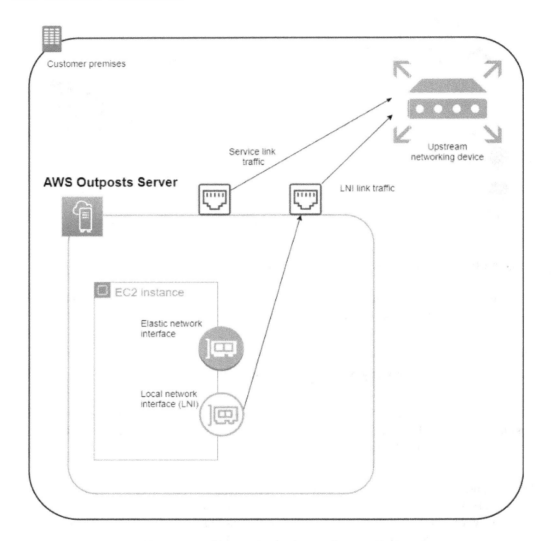

Figure 2.16 – Outposts servers' network ports scheme

The routing and logical segmentation of all this traffic falls outside the scope of this book. It suffices to say that all the routing is done using BGP, and network layer 2 segmentation is done using VLANs. We will provide references for advanced topics in *Chapter 8, Architecture References*, but here we provide a diagram depicting this segmentation for reference.

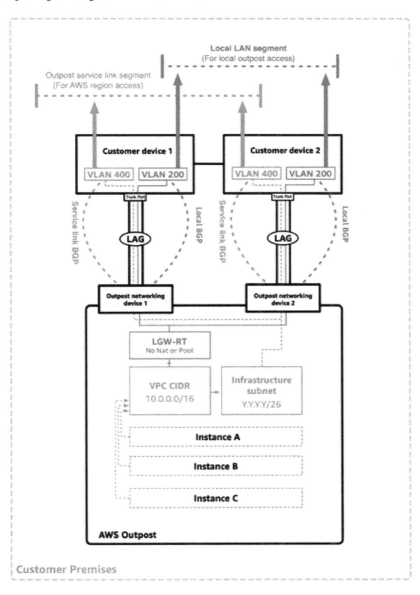

Figure 2.17 – Outposts use BGP for dynamic routing and VLANs for logical traffic segregation

Now we have AWS Outposts alive and communicating with its parent Region and with the customer's local network. Time to take it for a spin, put its capabilities to good use, and see what it can do. In the next section, we will talk about the services and features available on AWS Outposts.

Capabilities – services and features

It's easy to be amazed by the engineering and technology in AWS Outposts. As technicians, we love to spend most of our time talking about the hardware components and hypervisor bits that make up solutions such as AWS Outposts. But ultimately applications are built using the services running on top of that hardware, and that is what we will be talking about in the remainder of this chapter.

The first question that comes to minds with all the knowledge we already have about services is, what AWS services and capabilities are available on AWS Outposts?

In all fairness, it is unreasonable to expect that a single rack would be capable of running the full portfolio of AWS services. Even if we were to aggregate several racks logically to resemble something like an AWS Local Zone, that expectation would still not be realistic. AWS had to decide which specific services would be available on Outposts by taking into account customer demand, strategy, and use cases.

But not only that – there was also the challenge of engineering an AWS service that usually runs in an AWS Region to run on an AWS Outposts unit. The Service Link still provides a lifeline of communication with the AWS Region, but the AWS engineers decided to *divide and conquer*. Each service was split into software bits that were capable of operating on the edge closer to the customer's local networks to address the requirement for low latency, and the remaining components live in the AWS Region and interoperate via the Service Link.

The result of this decoupling is the architectural decision to run the service control plane in the Region, while the latency-sensitive parts are run inside the rack close to the customer's local workloads. Services running inside AWS Outposts have been optimized for operation across this split between the control plane and data plane.

This strategy certainly opened the doors for many AWS services to be made available in Outposts, but there are still other things to consider. For example, some services are multi-AZ by design while AWS Outposts is single-AZ. A solution that provides the levels of availability and durability of services such as Amazon Aurora inside AWS Outposts is still underway, and we have learned not to doubt AWS's engineering capabilities to pull off such achievements.

AWS Outposts rack comes in two flavors: one is the substrate on which to run AWS services on top of, and the other is dubbed *VMware Cloud on AWS Outposts*, dedicated to running VMware workloads utilizing the **Software-Defined Data Center (SDDC)** solution.

VMware Cloud on AWS Outposts runs on bare-metal AWS Outposts infrastructure. Similar to AWS Outposts, this offering is a fully managed service, where VMware is the single point of contact for support while AWS provides the underlying hardware, proactively monitors its components, and replaces them in case of failure.

This offering is targeted at customers currently leveraging VMware products in their data centers because they likely have the required skillset in-house along with familiarity with the product. Mass migrations can be performed using VMware HCX, which also offers the ability to extend capacity to VMware Cloud on AWS running in an AWS Region.

This solution operates similarly to the AWS Outposts rack, extending an underlying VPC that exists in an AWS account managed by VMware. Using the same principle as with the rack, one subnet can be created in the VPC linked to an Outposts and any VMs associated with this subnet will be launched inside the Outposts rack. This whole process is transparent to the customer, the rack will be shipped with the SDDC stack and after activation, you just need to start deploying your workloads on it.

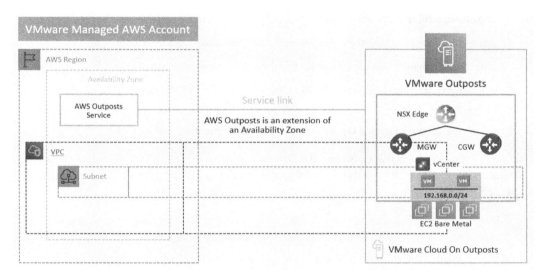

Figure 2.18 – VMware Cloud on AWS Outposts architecture

The brightest star is the Outposts rack flavor that brings AWS services on-premises. AWS Outposts is sold in distinct SKUs, which translate to specific configurations. The best scenario is to find a pre-validated configuration that meets your requirements. This will simplify the process, as you can place your order more quickly. The order process and pricing will be discussed in more detail in *Chapter 3, Pricing, Ordering, and Installation*.

Compute and storage

Let's talk about *compute* and *storage* options. AWS Outposts will always have EC2 and Amazon **Elastic Block Store** (**EBS**) services built into the rack. Customers cannot order less than the minimum configuration of one Droplet (individual server) and 11 TB of EBS storage.

For compute, currently, AWS Outposts only offers Intel instances with support for Graviton instances coming soon. You can select between five AWS EC2 instance types, each one purposely tailored to run certain types of workloads. One important concept to understand in order to get the right match between instance type and workload characteristics is the **vCPU to memory** (**vCPU:memory**) ratio.

The rationale behind this calculation is simple: *compute-intensive* workloads need more vCPU and less memory, while *memory-intensive* workloads are exactly the opposite. In the middle between these two extremes, we identify the *general-purpose* instance type, which offers a more balanced option and is the best option if you are uncertain about your workload profile. Other use cases include *graphics optimized* and *I/O optimized*.

- **Compute-optimized instance types**: C5 family, vCPU:memory ratio is 1:2

- **General-purpose instance types**: M5 family, vCPU:memory ratio is 1:4

- **Memory-optimized instance types**: R5 family, vCPU:memory ratio is 1:8

- **Graphics-optimized instance types**: G4dn family powered by NVIDIA T4 GPUs, vCPU:memory ratio is 1:4 + GPU.

- **Storage-optimized instance types**: I3en family, vCPU:memory ratio is 1:8

The base EC2 instance size is the *xlarge* and the Droplet (physical server unit) size is *24xlarge*, except for accelerated compute (GPU-enabled) units where the size is *12xlarge*. In the following list, we correlate each Droplet capacity with the corresponding vCPU:memory ratio:

- **C5**: Has a ratio of 1:2. The xlarge size corresponds to 4 vCPUs and 8 GiB of memory. Each droplet provides 96 vCPUs (24 x 4) and 192 GiB of memory (24 x 8).

- **M5**: Has a ratio of 1:4. The xlarge size corresponds to 4 vCPUs and 16 GiB of memory. Each droplet provides 96 vCPUs (24 x 4) and 384 GiB of memory (24 x 16).

- **R5**: Has a ratio of 1:8. The xlarge size corresponds to 4 vCPUs and 32 GiB of memory. Each droplet provides 96 vCPUs (24 x 4) and 768 GiB of memory (24 x 32).

- **G4dn**: Has a ratio of 1:4. The xlarge size corresponds to 4 vCPUs, 16 GiB of memory, and 1 GPU. Each droplet provides 48 vCPUs (12 x 4), 192 GiB of memory (12 x 16), and 4 GPUs.

- **I3en**: Has a ratio of 1:8. The xlarge size corresponds to 4 vCPUs and 32 GiB of memory. Each droplet provides 96 vCPUs (24 x 4) and 768 GiB of memory (24 x 32).

For block storage, we have the traditional option to use Amazon EBS, which is logically attached to the instance via an IP network and consumes network bandwidth. This option persists the storage across the stop-and-start cycles of your EC2 instance. Only **General Purpose SSD** (**GP2**) is supported.

To address the requirement for extremely high IOPS there is the option to use NVMe-based SSD storage. This component is physically connected to the host server (instance store) but the volumes created out of it will not be persisted across stop-and-start cycles; however, it will be persistent across reboots. If you stop your instance, the contents stored in these volumes will be gone. This option can be identified by the letter *d* appended to each instance type:

- **Compute-optimized instance types**: You can select between C5 (backed by EBS) and C5d (backed by NVMe-based SSD storage)

- **General-purpose instance types**: You can select between M5 (backed by EBS) and M5d (backed by NVMe-based SSD storage)

- **Memory-optimized instance types**: You can select between R5 (backed by EBS) and R5d (backed by NVMe-based SSD storage)

- **Graphics- and Storage-optimized instance types**: You have only NVMe-based SSD storage options

If you do not find a suitable prevalidated SKU (resource ID) to address your needs, you have the option to request a custom SKU. To create this configuration, you need to specify the amounts of compute, memory, and storage required, and the individual elements will be combined to achieve the desired capacity. Configurations are created based on sets of options where you should pick the most appropriate value:

- EBS' available tiers are 11 TB, 33 TB, and 55 TB. Larger storage capacities need to be validated by product engineering.

- A customized mix of EC2-instance families and types also needs to be validated by product engineering. As a rule of thumb, avoid mixing several instance families and specifying a wide variety of instance sizes. Ideally, create a baseline of three instance sizes and try to fit your workloads within these three sizes. This will significantly help the resilience of the design.

Containers

Now let's talk *containers*. You can request your AWS Outposts unit to be able to run Docker-based Amazon **Elastic Container Service** (**ECS**) and Kubernetes-based Amazon **Elastic Kubernetes Service** (**EKS**). Regardless of the flavor you choose, there are important characteristics of these services running on AWS Outposts that need to be well understood:

- The *control plane* lives in the Outposts parent Region and the *containers* (ECS container instances and EKS worker nodes) are executed inside the rack. The control plane generates traffic traversing the Service Link for ECS (task definitions) and EKS (pod specs). Similarly, metrics and CloudWatch logs transit to the AWS Region.

- Application traffic can be restricted to Outposts and the local network.

- If you have a disconnected Service Link, containers will continue to run but not indefinitely and it is difficult to specify for how long they will remain stable. Mutation calls (changes) will not be possible.

- At the time of writing, you cannot use AWS Fargate to orchestrate containers deployed on AWS Outposts.

- Container architectures and designs that leverage AWS services living in the region must consider the latency imposed by the Service Link. We can cite AWS **Identity and Access Management (IAM)**, **Amazon Elastic Container Registry (Amazon ECR)**, Network Load Balancer, Classic Load Balancer, and Amazon Route 53.

- If you use **Elastic Container Registry (ECR)**, note that it has a dependency on S3 in the Region. Container images are pulled from S3 via Service Link and cached locally on AWS Outposts.

- The **Application Load Balancer (ALB)** is available to run on Outposts alongside clusters and worker nodes. If you need functionality similar to that provided by **Network and Classic Load Balancers (NLB/CLB)**, you can deploy third-party solutions on premises (for example, F5 BIG IP or NetScaler).

- You can't deploy ECS or EKS anywhere on Outposts. These technologies can leverage bare-metal servers, VMs, and even small compute devices such as the Raspberry Pi.

Aside from these limitations, you can leverage the same APIs and tools to deploy containerized applications on AWS Outposts to communicate with your on-premises networks and address use cases such as low latency requirements.

Here is a diagram showing the architecture of Amazon ECS and its components.

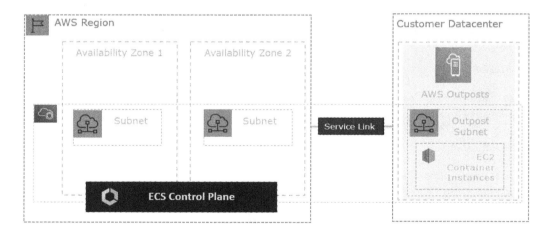

Figure 2.19 – ECS control plane and data plane architecture

Now, let's contrast that with the following diagram showing the architecture of Amazon EKS and its components, where we can identify the addition of EKS master nodes that live in the Region.

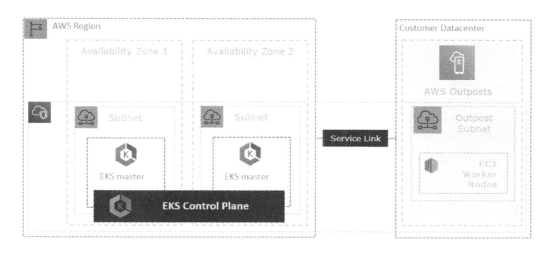

Figure 2.20 – EKS control plane and data plane architecture

Databases

Time to talk *databases*. AWS Outposts offers the ability to leverage Amazon **Relational Database Service (RDS)**. This fully managed service allows you to deploy and manage various database engines easily and in an automated fashion. To do this, you use the same AWS console, APIs, and tools you typically use when working in the Region.

Currently, Amazon RDS on Outposts supports the Microsoft SQL Server, MySQL, and PostgreSQL database engines, and although not all versions and features are available for these engines, you can expect this scope to expand in the future. With RDS on Outposts, you can address use cases for applications that need access to local databases with low latency.

Data residency requirements can be achieved to a certain extent but guardrails and policies must be in place to ensure these criteria are met. One thing to be mindful of is that local snapshots of RDS databases by default are stored in the Region. Local snapshots of RDS databases are supported but you need to have S3 on Outposts.

The distinct characteristics of this service running on AWS Outposts are the following:

- RDS instances are only available on *M5* and *R5* Droplets with sizes going up from *large* to *24xlarge*.

- The only storage option available is *general-purpose SSD* and you cannot change the allocated storage.

- It neither supports *multi-AZ deployments* nor *read replicas*.

- At this time, you cannot leverage *placement groups* with RDS on Outposts.

- *Managed replication* is not available. However, *customer-managed replication* can be configured using DB-native tools or third-party tools.

- RDS on Outposts can use a local CoIP to provide local connectivity.

- You must design your architecture to withstand Service Link disconnections. A local secondary DNS server can help with name resolution during interruptions. Automatic backups will not occur if connectivity is not established; moreover, DB instances are not automatically replaced (via auto-recovery) until connectivity is restored.

- You cannot restore a DB instance from Amazon S3 but you can export DB snapshot data to S3 (the engines supported are MySQL and PostgreSQL).

Another service available is Amazon ElastiCache for architectures that require an in-memory data store to support real-time applications with sub-millisecond latency. The service is fully managed and compatible with the *Redis* and *Memcached* engines. Caching strategies can address use cases such as session stores, gaming, geospatial services, real-time analytics, and message queuing.

The distinct characteristics of this service running on AWS Outposts are as follows:

- ElastiCache is only available on *M5* and *R5* Droplets

- *Multi-AZ* (cross-Outposts replication is not supported)

- Local snapshots are not supported; automated backups are stored in the AWS Region

- Live migration is supported using the *Redis* engine

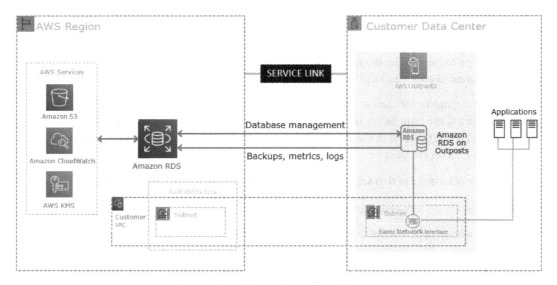

Figure 2.21 – RDS on Outposts architecture

Big data analytics capability is available on AWS Outposts with Amazon **Elastic Map Reduce (EMR)**. Amazon EMR leverages open source tools such as Apache Hadoop, Apache Hive, Apache Spark, and Presto as managed clusters to analyze and process large datasets. You can also run Amazon EMR container jobs on top of EKS clusters running on Outposts.

The distinct characteristics of this service running on AWS Outposts are as follows:

- You cannot use Spot instances in Amazon EMR running on AWS Outposts
- The only storage option available is *general-purpose SSD*
- You can only leverage Amazon S3 for Amazon EMR on Outposts; S3 on Outposts is not supported
- You must be running Amazon EMR version 5.28 and above

S3

Finally, let us talk about S3 on Outposts. This was one of the most anticipated features of AWS Outposts, but when it was launched in December 2019, it was not yet ready for primetime. The availability of S3 on AWS Outposts was announced in September 2020.

It is beneficial to dive a little deeper to understand how this service was adapted to work on AWS Outposts. Let us explore the service running in the Region and how it is positioned.

Consider the following statement: the total volume of data and the number of objects you can store in S3 are unlimited. If we think in terms of Regions, of course, we must not take the word *unlimited* literally. This is better understood as *virtually unlimited*, meaning in practical terms AWS will always have space available if you need it. They will not say how much storage is allocated to S3, AWS just says *don't worry about it*.

Of course, the same does not apply to S3 on Outposts; storage is limited to 380 TB using AWS native solutions. You can push this limit via validated third-party providers such as Cloudian for AWS Outposts (`https://cloudian.com/aws/`).

Amazon S3 in the Region offers several storage classes: *S3 Intelligent-Tiering, S3 Standard, S3 Standard-Infrequent Access, S3 One Zone-Infrequent Access, Amazon S3 Glacier Instant Retrieval, Amazon S3 Glacier Flexible Retrieval* (formerly S3 Glacier), and *Amazon S3 Glacier Deep Archive*. S3 on Outposts offers one single storage class called *S3 Outposts*, which achieves the same levels of durability and redundancy of the *S3 Standard* storage class in the region, by storing data across multiple devices and servers.

In the Region, if you want to use encryption to store objects in S3 you can choose between *server-side encryption* (using *SSE-KMS* or *SSE-S3*) and *client-side encryption* (*SSE-C*). S3 on Outposts encrypts your data by default using *SSE-S3* as well as giving you the option to encrypt with your own encryption keys (*SSE-C*).

In the Region, S3 endpoints are global bucket names automatically created for you. To store objects in buckets created in S3 on Outposts, you need to create an *S3 Access Point* to associate your VPC with one bucket created inside AWS Outposts and next create an *S3 API endpoint* to associate your subnet with the logical Outposts ID.

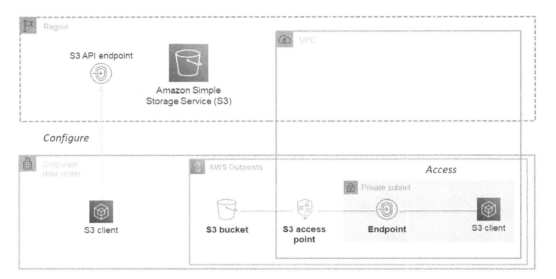

Figure 2.22 – S3 on Outposts architecture

The distinct characteristics of this service running on AWS Outposts are as follows:

- Data is always stored on Outposts: object data, system data, and user metadata.

- Buckets are created and managed in the Outposts' home Region.

- Telemetry and management data are available in the home Region.

- S3 on Outposts buckets are not accessible outside their connected VPCs. You can create and access S3 objects on Outposts directly from your on-premises network using *direct access*, which basically comprises creating S3 Access Points inside Outposts utilizing CoIP addresses.

- *S3 Object Ownership* is always enabled.

- The *Block public access* feature is always enabled.

- Local S3 requests will fail if the AWS Outposts unit is disconnected from the Region because of the dependency with *IAM*.

- Because storage capacity is finite, if you run out of storage the S3 API response will be **HTTP 507 Insufficient Storage**.

Outposts servers are obviously much more restricted in terms of the services they can run; however, they have the option to come with AWS Graviton2 processors based on ARM architecture.

> **Note**
>
> AWS Outposts servers only support local storage and not EBS volumes; therefore, you always have to be mindful of your backup and HA strategy for instances running inside these devices.

The Intel flavor (2U) has two options: *c6id.16xlarge* and *c6id.32xlarge*.

The *c6id.16xlarge* model (64 vCPUs and 128 GB of memory) supports six instance sizes with the exception of the largest one, supported on the *c6id.32xlarge* model (128 vCPUs and 256 GB of memory), as follows:

Instance Name	vCPUs	Memory	Local Storage
c6id.large	2	4 GiB	118 GB
c6id.xlarge	4	8 GiB	237 GB
c6id.2xlarge	8	16 GiB	474 GB
c6id.4xlarge	16	32 GiB	950 GB
c6id.8xlarge	32	64 GiB	1.9 TB
c6id.16xlarge	64	128 GiB	3.8 TB
c6id.32xlarge	128	256 GiB	7.6 TB

Figure 2.23 – Instances available on Outposts server 2U

The Graviton2 offering (1U) has only one option: *c6gd.16xlarge*.

This model supports six instance sizes, as follows:

Instance Name	vCPUs	Memory	Local Storage
c6gd.large	2	4 GiB	118 GB
c6gd.xlarge	4	8 GiB	237 GB
c6gd.2xlarge	8	16 GiB	474 GB
c6gd.4xlarge	16	32 GiB	950 GB
c6gd.8xlarge	32	64 GiB	1.9 TB
c6gd.16xlarge	64	128 GiB	3.8 TB

Figure 2.24 – Instances available on Outposts Server 1U

Besides the ability to run the traditional Amazon **Elastic Compute Cloud** (**EC2**), you also have available Amazon **Elastic Container Service** (**ECS**) clusters to run on *Outposts Servers*.

Additionally, all members of the AWS Outposts family are capable of running the *AWS App Mesh Envoy* proxy, *AWS IoT Greengrass*, and *Amazon SageMaker Neo*.

This was a long section because AWS Outposts is a very powerful platform with a lot of services and capabilities available to build compelling applications and solutions for edge-bound use cases. In the next section, we will briefly discuss some connectivity-specific elements and configuration options.

Connectivity elements – cables and connectors

This section is brief. It will prepare you to connect your AWS Outposts rack with your on-premises networking devices. AWS opted to keep the available options simple, which reduces the number of choices to be made. Performance and reliability are the criteria when it comes to connector and media options and AWS is strict with these criteria.

AWS has done an excellent job describing Outposts requirements at `https://docs.aws.amazon.com/outposts/latest/userguide/outposts-requirements.html`. We will reproduce some of the tables while exploring the topics:

- *Network speeds*: You can choose between 1 GBps, 10 GBps, 40 GBps, or 100 GBps.

> **Important**
>
> If you take the route of 1 GBps, the **Outposts Networking Device** (**OND**) will only support this link speed. The other OND available for Outposts supports the speeds of 10, 40, and 100 GBps but exclusively. If you select 10 GBps uplinks, only this link speed will be configured.

- *Number of uplinks*: Depending on the network speed selected, the maximum number of uplinks available will vary as follows:

  - *1 GBps*: 1, 2, 4, 6, or 8 uplinks

  - *10 GBps*: 1, 2, 4, 8, 12, or 16 uplinks

  - *40 GBps or 100 GBps*: 1, 2, or 4 uplinks

- *Media type*: Only fiber optic is available – there are no copper wire options. You can use single-mode or multi-mode fiber with Lucent Connectors.

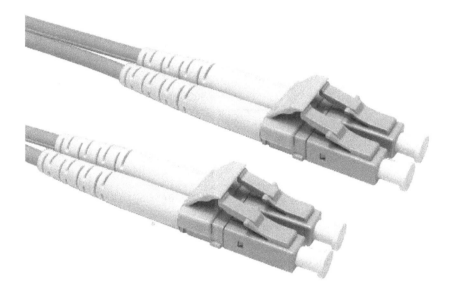

Figure 2.25 – Lucent Connectors

There are several optic standards available, as seen in the following table:

Uplink speed	Fiber type	Optical standard
1 Gbps	SMF	– 1000Base-LX
1 Gbps	MMF	– 1000Base-SX
10 Gbps	SMF	– 10GBASE-IR – 10GBASE-LR
10 Gbps	MMF	– 10GBASE-SR
40 Gbps	SMF	– 40GBASE-IR4 (LR4L) – 40GBASE-LR4
40 Gbps	MMF	– 40GBASE-ESR4 – 40GBASE-SR4
100 Gbps	SMF	– 100G PSM4 MSA – 100GBASE-CWDM4 – 100GBASE-LR4
100 Gbps	MMF	– 100GBASE-SR4

Figure 2.26 – Outposts Networking Device-supported optical standards

- *Link aggregation*: Uplinks can be aggregated using dynamic LAG with LACP. Cross-device (between Outposts Networking Devices) LACP is not supported.
- *Demarcation point*: Outposts defines the physical demarcation point as the fiber patch panel on Outposts. Beyond that point, everything is the responsibility of the customer, including the provision of any fiber cables that run between the customer's networking devices and Outposts' physical patch panel.

Figure 2.27 – Outposts' patch panel

To close this section, let's consider one strategic piece of information about selecting the upstream switches to which your AWS Outposts unit will be connected: there is no price difference whether you select the Outposts Networking Device capable of delivering 10, 40, or 100 GBps network speeds.

You already paid for 100-GBps ports: if you connect to a 10-GBps upstream device you are wasting considerable bandwidth. It would be wise to invest in upgrading your upstream devices to utilize the full potential of your Outposts networking potential instead of saving some money at expense of performance.

The performance of your AWS Outposts unit when interacting with your corporate network and the parent AWS Regions is completely dependent on good connectivity and generous bandwidth. In this section, you explored the full range of options available to connect your rack.

Summary

That's a wrap! It was a long journey but now you can be confident about your AWS Outposts unit. You know how it looks at a glance and how it's built structurally, along with the requirements and options to power it up, get set, and go!

You have a deep understanding of what's inside when you pop the hood, and what you can do with it. An in-depth view of each service and its limitations gave you all the knowledge necessary to build solutions and architect with confidence on Outposts. The fundamental role of the network in the performance of your Outposts unit was also highlighted. Its capabilities certainly stand out from the crowd.

Now you are undoubtedly anxious to get your hands dirty. In the next chapter, we will discuss the AWS Outposts installation and show you how to configure it so you can begin building your hybrid solutions.

3
Pricing, Ordering, and Installation

After having gained knowledge about the various aspects of AWS Outposts, including its capabilities, design, engineering, and connectivity, you should feel confident in architecting solutions that are capable of running on AWS services both on Outposts and in an AWS Region. After the architecture and use case passes all checks and balances and proves to be valid, we feel the drive to jump into the AWS Management Console and order a rack.

However, as with most of the things related to AWS Outposts, the ordering process is not similar to ordering traditional hardware. AWS would like to ensure that the customer knows all the requirements and implications of ordering AWS Outposts equipment.

Going through the various steps of the ordering process, diligently answering the questions, and placing an order is just a manifestation of an intention to purchase an AWS Outposts rack. However, for AWS Outposts to ship to the customer, several other steps need to be fulfilled, and this chapter intends to help you with that process.

This chapter will cover the following:

- The pricing structure of AWS Outposts and where to find prices and configurations
- Communication elements – networking
- Capabilities – services and features

AWS Outposts pricing options

You are now qualified to describe the engineering aspects of AWS Outposts, use cases, capabilities, and what AWS services it can run. We also touched upon some of its requirements and limitations. AWS Outposts is a fascinating platform that enables a myriad of hybrid architectures. Ready to get one for your data center? Well, let's see what that means financially.

There is no option to buy and have an AWS Outposts rack or server registered as an asset on your books. With Outposts, you are always purchasing capacity and not hardware – it can only run AWS services and only AWS can perform the configuration, maintenance, and provide spares. Your only option is to purchase the rack for a fixed term.

If you access the AWS pricing page for an AWS Outposts rack at this URL (`https://aws.amazon.com/outposts/rack/pricing/`), the options shown here are for a three-year term. Other terms are available, but those can be obtained by contacting AWS directly. The **Resource ID** column shown on this page corresponds to all the available pre-validated SKUs.

Each SKU represents a unique configuration capable of running a finite number of instances. The instance types available to be launched on a given SKU are dependent on the underlying server hardware present in the configuration. This information is not provided on the pricing page, but we can get more details using the AWS Management Console. The following is a step-by-step walkthrough – it requires an active AWS account:

1. Log on to the AWS Management Console at `https://console.aws.amazon.com/`:

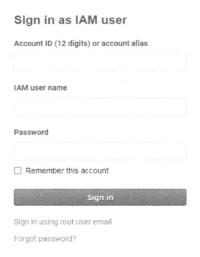

Figure 3.1 – The AWS Console login page

2.   Enter Outposts in the search bar and be cautious to make sure you select the desired region:

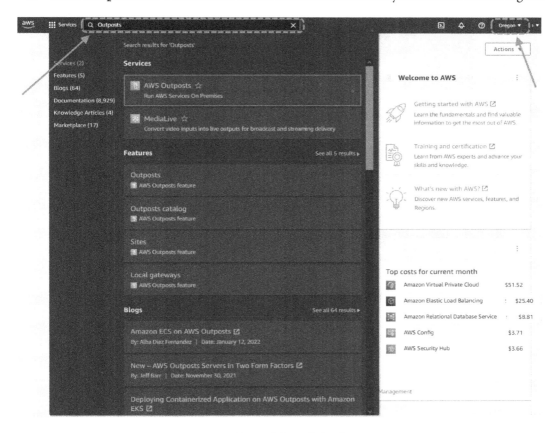

Figure 3.2 – The AWS Console landing page

3. Click **Browse catalog**:

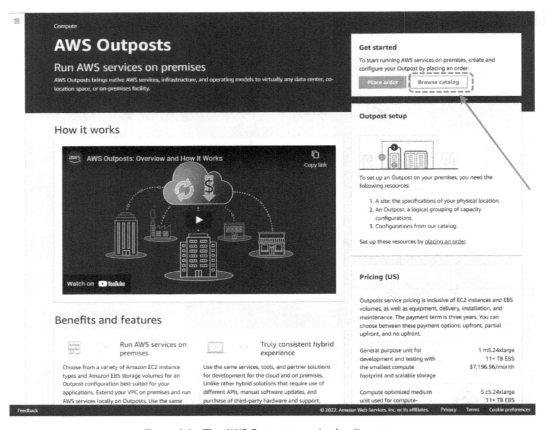

Figure 3.3 – The AWS Outposts service landing page

4. On the **Outposts catalog** page, you have access to the configurations of Outposts servers or racks. On this page, you will find the specs of each pre-validated SKU:

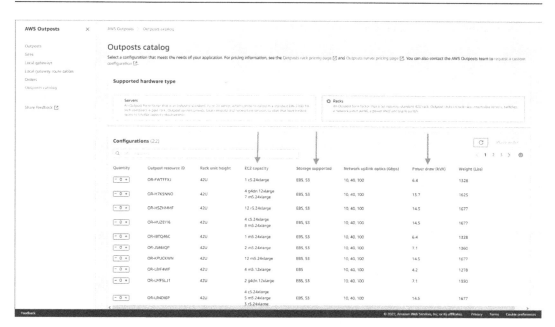

Figure 3.4 – The AWS Outposts racks catalog

The **EC2 Capacity** column tells you the number of EC2 instances supported by the hardware alongside the corresponding instance family and type.

The **Storage supported** column tells you whether a given SKU only supports EBS or EBS and S3. The **Power draw** column gives the rack consumption, but we must remember that AWS will require either 5, 10, or 15 kVA. Therefore, to meet the AWS requirements for power, the value to be considered should be the **Power draw** value rounded up to the next value within these three options.

As we mentioned, custom SKUs can be worked out with the AWS team and each one will correspond to a unique configuration. When custom SKUs are requested, they will be shown in this section once approved by the AWS Outposts product engineering team, but only for specific accounts that must be defined by the customer. You can submit your request at this URL: `https://pages.awscloud.com/AWS-Outposts-capacity-configuration.html`.

Here is a snapshot of the catalog page for Outposts servers:

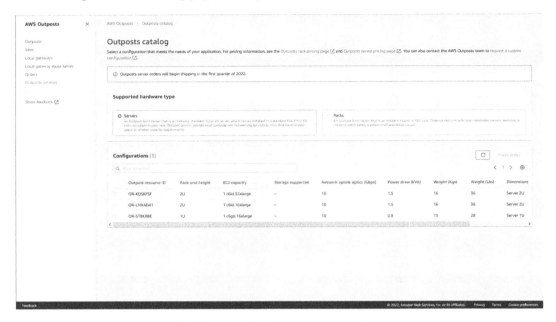

Figure 3.5 – The Outposts servers catalog

Remember, Enterprise Support is mandatory and you must be enrolled for it with AWS Support for the duration of your AWS Outposts usage. You can find more information about Enterprise Support at this URL: `https://aws.amazon.com/premiumsupport/pricing/`.

What is included in the price of an *Outposts rack*?

- Rack delivery and installation, anywhere in the country

- Servicing (maintenance, software patches, and upgrades)

- Decommissioning and return shipment costs at the end of the contractual term

What is included in the price of *Outposts servers*?

- Server delivery, anywhere in the country

- Service (maintenance, software patches, and upgrades)

Now, we will examine what is *not included* in the price.

For an *Outposts rack*, this is as follows:

- Storage capacity is not included and it is consumed by EBS, EBS Snapshots, and S3 on Outposts. You cannot order a rack without EBS – therefore, at least 11 TB of storage will be added to the price seen on the product page. Once you place an order, you will receive a contact from an AWS sales representative to work through all these details and a final quote will be provided for customer approval.

For *Outposts servers*, this is as follows:

- Installation (by following along with the AWS documentation) is the responsibility of the customer
- Removal and shipping the unit back to AWS (at the end of the term) is the responsibility of the customer

For both, *racks and servers*:

- **Operating System (OS)** charges. These are applied based on usage per instance.
- AWS services running locally are billed based on their respective pricing – there will be further details on this later in this chapter.
- Solutions available in AWS Marketplace are charged at the same price as in the parent AWS Region.
- Data transfer charges. There are no charges for data transfers across the local network (it happens via the *LGW*), but egress or ingress traffic from and to the Region may be charged.

AWS services running on AWS Outposts are not covered by the same commitments or agreements made by AWS for the same services running in the Region. In the Region, AWS operates and maintains all the necessary infrastructure necessary to support these services, such as power, networking, and security. However, in the case of AWS Outposts, the responsibilities are distributed between the customer and AWS based on the AWS shared responsibility model.

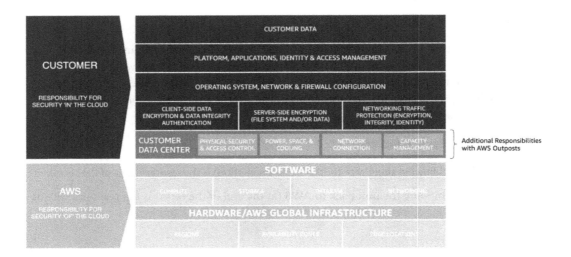

Figure 3.6 – The AWS shared responsibility model updated for AWS Outposts

Let's now understand the pricing scheme for the services running on the rack. When you order the rack, there will be a separate item on the bill that refers to the rack SKU itself. The services will be charged in their respective sections on the AWS bill.

Pricing for services running on Outposts

Just as AWS service pricing varies by region, AWS Outposts service pricing varies based on the parent region associated with AWS Outposts during the order process. However, the variance in pricing does *not* apply to AWS Outposts servers since they only run EC2 and do not have EBS for persistent storage:

- **Amazon EC2**: This service is available on both racks and servers. When a customer purchases AWS Outposts, the total capacity is allocated to the customer for consumption. Also, customers are not additionally billed per EC2 instance. However, OS charges based on usage are billed to the customer to cover the licensing fees whenever applicable (for example, Windows).

- **Amazon EBS**: Prices vary per region according to specific tiers on a *per GB-month* basis. AWS defines three base tiers: 11, 33, and 55 TB. Above 55 TB, you can scale in increments of 15 TB for 70, 85, and 100 TB. Beyond 100 TB, you scale in increments of 20 TB up to 1 PB.

  The EBS rate will remain constant based on the initial tier you choose. Therefore, the best strategy is to select a higher tier to benefit from a lower rate.

  EBS volume snapshots are stored in the AWS parent region by default, but if you have S3 on Outposts, you can take snapshots using this storage capacity. EBS local snapshots on Outposts are priced separately according to the parent region where Outposts is anchored on a *per GB-month* basis.

- **Amazon S3**: Prices vary per region according to specific tiers on a *per GB-month* basis. AWS defines tiers for 26, 48, 96, 240, or 380 TB. You are not charged for S3 API operations against buckets and objects on an Outposts rack. There is no ingress cost for data going into a S3 bucket on an Outposts rack and no egress cost for getting data from a S3 bucket to an Outposts rack to the local network via the LGW.

Services running on AWS Outposts may utilize EC2 instances and EBS storage, which is the case for Amazon RDS, Amazon EMR, Amazon EKS, and Amazon ECS. The costs associated with AWS services such as Amazon RDS, Amazon EMR, Amazon EKS, and Amazon ECS on AWS Outposts exclude the costs associated with the underlying EC2 instance and EBS storage:

- **Amazon RDS**: The DB instances can be selected between M5 and R5 instance types and with six sizes available: large, xlarge, 2xlarge, 4xlarge, 12xlarge, and 24xlarge. Prices are billed hourly per instance – fractions of an hour will be rounded up and considered a full hour. The price details can be found here: `https://aws.amazon.com/rds/outposts/pricing/`.

  You can create a price estimation using AWS Pricing Calculator for RDS on AWS Outposts, found at this URL: `https://calculator.aws/#/createCalculator/RDSOutposts`.

- **Amazon EMR**: Priced similar to Amazon EMR running in the AWS Region, excluding EC2 and EBS costs. Customers are bound to the instance families and types available on the Outposts rack. The pricing details can be found here: `https://aws.amazon.com/emr/pricing/`.

- **Amazon ElastiCache**: ElastiCache for Outposts only supports M5 and R5 family nodes. Prices vary per region and can be found at this URL: `https://aws.amazon.com/elasticache/pricing/`.

- **Amazon EKS**: Priced similar to Amazon EKS running in the AWS Region, excludes EC2 and EBS costs. An EKS cluster is deployed in the AWS Region and the worker nodes will be deployed and consume the AWS Outposts rack capacity. The pricing details can be found here: `https://aws.amazon.com/eks/pricing/`.

- **Amazon ECS**: ECS costs are purely based on AWS resource usage (for example, Amazon EC2 instances or Amazon EBS volumes) – customers don't pay for the service. When deployed on AWS Outposts, ECS architecture uses the same strategy employed by EKS, with the control plane running in the cloud while the container instances are running on the Outposts compute capacity, which is available to you at no additional charge.

- **Application Load Balancer**: Application Load Balancer pricing is based on service, not capacity. The **Load Balancer Capacity Unit** (**LCU**) price is zero and the service is billed per hour. Pricing varies per region and further details can be found here: `https://aws.amazon.com/elasticloadbalancing/pricing/`.

- **Data transfer charges**: As a rule of thumb, data leaving AWS Outposts (to either the LGW or AWS Region via a service link) is not charged. Data transfer charges within the same AWS Region for intra-AZ, inter-AZ, and VPC network traffic remain the same when applied to data packets entering AWS Outposts originating from the parent **Availability Zone (AZ)**. These are described on the EC2 pricing page here: `https://aws.amazon.com/ec2/pricing/on-demand/#Data_Transfer`.

When AWS **Direct Connect (DX)** provides the connectivity back to the AWS anchor region, ensure that the ingress and egress charges associated with AWS DX are considered. The costs can be found here: `https://aws.amazon.com/directconnect/pricing/`. Also, the following diagram depicts how the pricing works regarding network traffic (see the details at `https://docs.aws.amazon.com/whitepapers/latest/how-aws-pricing-works/aws-outposts.html`):

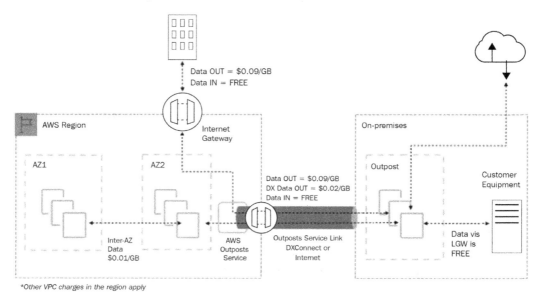

Figure 3.7 – An example of ingress or egress charges for AWS Outposts

We now have an in-depth understanding of how AWS Outposts is priced and how each service influences the cost of the solution over time. This is a key component to perform a comprehensive evaluation of **Total Cost of Ownership (TCO)** for AWS Outposts.

Let's remember that AWS Outposts does not compare to commodity hardware and an attempt to perform this side-by-side will never be an *apple-to-apple* comparison. The value that AWS Outposts brings to the table cannot be matched by traditional hardware offerings.Let's move ahead and see how we can place an order and have AWS Outposts arrive at our doorstep. Once again, if you think of traditional hardware and consider you know how this is going to work, more surprises are coming your way.

Placing an order

We expanded our background and now we also understand the pricing options and how it varies depending on the selected configurations and among the regions. We know what is included in the price and the terms.

We logged on to the AWS console to get familiar with the interface, navigated to the AWS Outposts service section, and browsed the Outposts catalog to discover the pre-validated SKU capabilities and requirements.

Time to finally place the order! Fire your browser, navigate to the AWS Management Console page, log on, and head to the AWS Outposts service landing page. If you prefer, you can access the Outposts landing page at this address: `https://console.aws.amazon.com/outposts/`. The **Place order** button is pretty visible and distinct, but let's not go there. In the top-left corner, we see an area with a picture that appears to be three stacked bars, also called the hamburger menu:

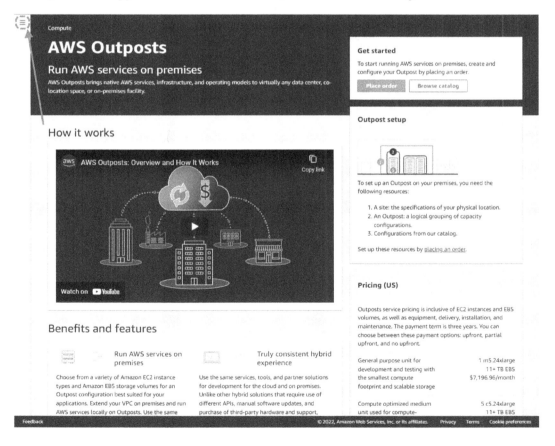

Figure 3.8 – The AWS Outposts service landing page

Clicking on that icon reveals the sidebar:

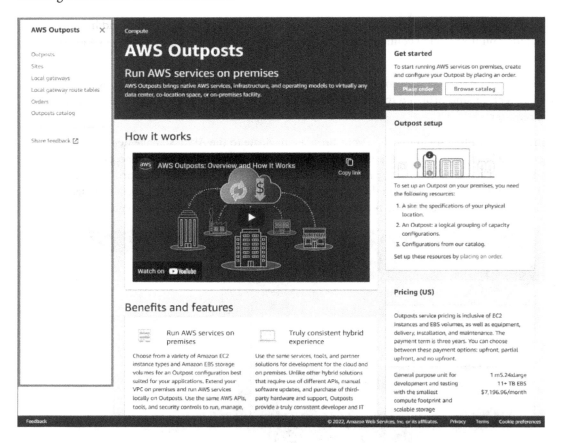

Figure 3.9 – The AWS Outposts service sidebar

To place an order, we need to understand how the process is structured. First, we need to create a *site*. Next, we create an *Outpost* associated with that site. Only then we can place an *order*.

A site is a physical location where you add capacity. Capacity is provided by physical Outposts racks or servers. You can create multiple sites and at the time of creation, you have to specify whether that particular site will receive capacity provided only by Outposts servers or both Outposts racks and servers.

The reason for that distinction relates to the requirements of each product. The site is the place where the rack or server exists physically – therefore, it needs to provide power, cooling, networking, and stable ground. The requirements for Outposts servers are very simple to provide compared to Outposts racks. If a given site can provide the requirements for the rack, it will certainly be capable of hosting Outposts servers.

After a site is created, it cannot be changed later. You can only change its *name* and *description* and not its capabilities. If something changes after the site has been created, you will have to create a new one. You can create multiple sites with the same address and different capabilities; a good example will be a data center complex with multiple buildings.

Let's take a look at some screenshots of the ordering process. Click on **Sites** in the left-hand pane:

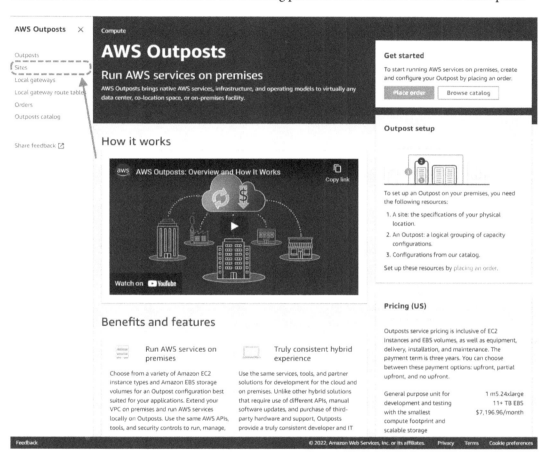

Figure 3.10 – The Sites menu in the AWS Outposts service

You will be taken to the **Sites** landing page. Next, click **Create site**:

Figure 3.11 – Creating a site on AWS Outposts

At **Create site**, we populate the needed information and click **Create site**. Common sections are **Site Information**, **Operating address**, and **Site notes**. When you select **Racks and servers**, there is an extra section called **Site details**.

Operating address is the section where the necessary power, network, and cooling infrastructure to support the rack is located:

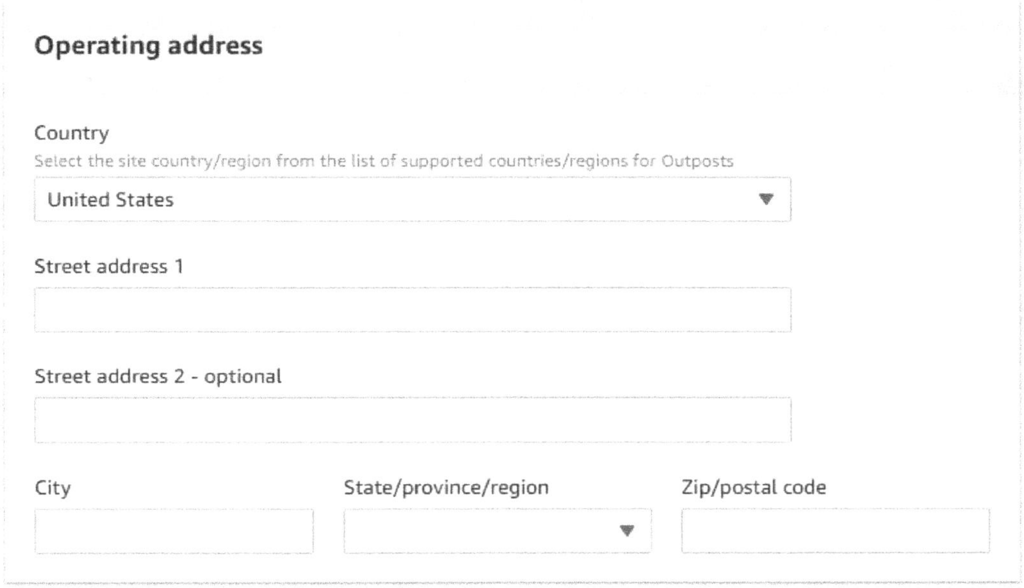

Operating address

Country

Select the site country/region from the list of supported countries/regions for Outposts

| United States ▼ |

Street address 1

Street address 2 - optional

| City | State/province/region | Zip/postal code |

Figure 3.12 – The Operating address section for Outposts

The **Site details** section outlines the specific configurations available at the operating address:

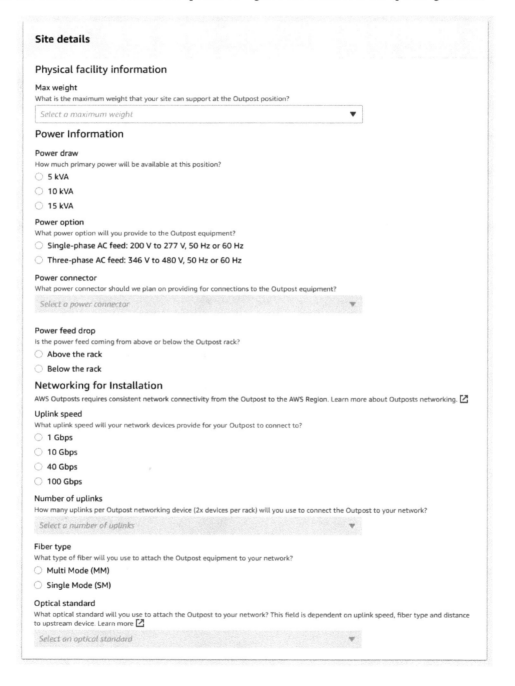

Figure 3.13 – The Sites details section for Outposts

The **Site Notes** section is available for you to highlight specific site characteristics and processes, which are relevant to assuring a smooth rack delivery:

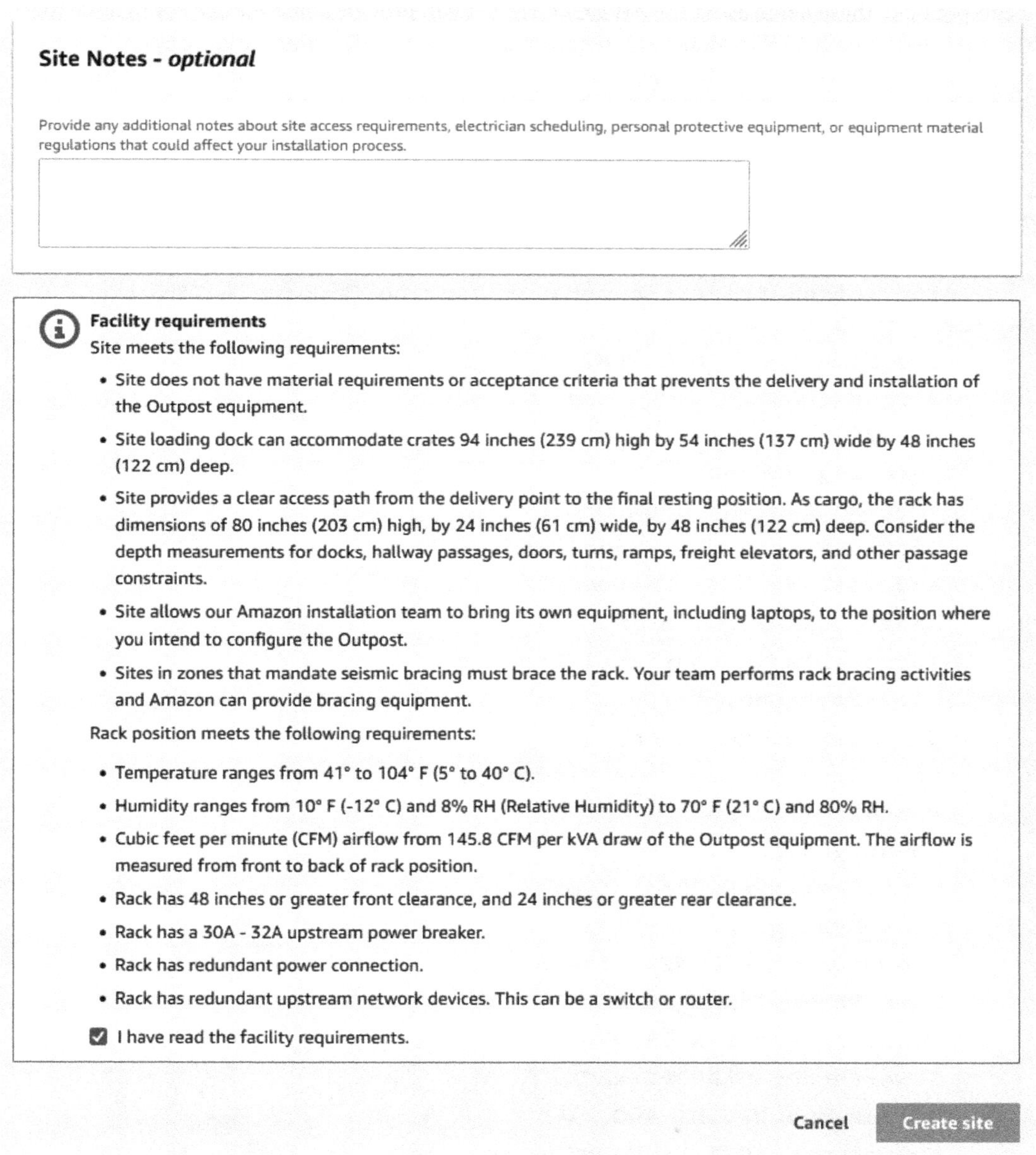

Site Notes - *optional*

Provide any additional notes about site access requirements, electrician scheduling, personal protective equipment, or equipment material regulations that could affect your installation process.

(i) Facility requirements
Site meets the following requirements:

- Site does not have material requirements or acceptance criteria that prevents the delivery and installation of the Outpost equipment.
- Site loading dock can accommodate crates 94 inches (239 cm) high by 54 inches (137 cm) wide by 48 inches (122 cm) deep.
- Site provides a clear access path from the delivery point to the final resting position. As cargo, the rack has dimensions of 80 inches (203 cm) high, by 24 inches (61 cm) wide, by 48 inches (122 cm) deep. Consider the depth measurements for docks, hallway passages, doors, turns, ramps, freight elevators, and other passage constraints.
- Site allows our Amazon installation team to bring its own equipment, including laptops, to the position where you intend to configure the Outpost.
- Sites in zones that mandate seismic bracing must brace the rack. Your team performs rack bracing activities and Amazon can provide bracing equipment.

Rack position meets the following requirements:

- Temperature ranges from 41° to 104° F (5° to 40° C).
- Humidity ranges from 10° F (-12° C) and 8% RH (Relative Humidity) to 70° F (21° C) and 80% RH.
- Cubic feet per minute (CFM) airflow from 145.8 CFM per kVA draw of the Outpost equipment. The airflow is measured from front to back of rack position.
- Rack has 48 inches or greater front clearance, and 24 inches or greater rear clearance.
- Rack has a 30A - 32A upstream power breaker.
- Rack has redundant power connection.
- Rack has redundant upstream network devices. This can be a switch or router.

☑ I have read the facility requirements.

Cancel   Create site

Figure 3.14 – Additional details and facility requirements for Outposts

All the available options for the **Site details** section can be found at this URL: `https://docs.aws.amazon.com/outposts/latest/userguide/outposts-requirements.html`.

You can populate these fields using the *AWS CLI*. Here, I will provide some sample commands that can be executed using AWS CloudShell. More information is available at this URL: `https://aws.amazon.com/cloudshell/`. The shell is a regular Linux BASH, and we will use the \ continuation character for readability. It helps to visualize lengthy commands.

To create a site, see the following:

```
aws outposts create-site \
    --name "DC02-FMC-RCK-OR03" \
    --description "Fictitious Manufacturing Company - IT
Datacenter 02 in Oregon" \
    --tags Name="OR03_DC02_FMT",\
        Status="LAB",\
        CC="2376" \
    --operating-address \
        "ContactName=Mr John Doe, \
        ContactPhoneNumber=+55 19 666-6666, \
        AddressLine1=777 SW Some Street, \
        AddressLine2=Tualatin, \
        AddressLine3=Industrial District, \
        City=Portland, \
        StateOrRegion=Oregon, \
        DistrictOrCounty=Some Vicinity, \
        PostalCode=97203, \
        CountryCode=US, \
        Municipality=DC02" \
    --rack-physical-properties \
        "PowerDrawKva=POWER_15_KVA, \
        PowerPhase=THREE_PHASE, \
        PowerConnector=AH530P7W, \
        PowerFeedDrop=ABOVE_RACK, \
        UplinkGbps=UPLINK_40G, \
        UplinkCount=UPLINK_COUNT_4, \
        FiberOpticCableType=MULTI_MODE, \
        OpticalStandard=OPTIC_40GBASE_SR, \
```

```
        MaximumSupportedWeightLbs=MAX_2000_LBS" \
    --notes "Loading Docks ~ Seismic requirements ~ Work hours";
```

The output will be something similar to what follows:

```
{
    "Site": {
        "SiteId": "os-0431a19da45a53dc0",
        "AccountId": "123456789012",
        "Name": "DC02-FMC-RCK-OR03",
        "Description": "Fictitious Manufacturing Company - IT
Datacenter 02 in Oregon",
        "Tags": {
            "CC": "2376",
            "Status": "LAB",
            "Name": "OR03_DC02_FMT"
        },
        "SiteArn": "arn:aws:outposts:us-west-
2:123456789012:site/os-0431a19da45a53dc0",
        "Notes": "Loading Docks ~ Seismic requirements ~ Work
hours",
        "OperatingAddressCountryCode": "US",
        "OperatingAddressStateOrRegion": "Oregon",
        "OperatingAddressCity": "Portland",
        "RackPhysicalProperties": {
            "PowerDrawKva": "POWER_15_KVA",
            "PowerPhase": "THREE_PHASE",
            "PowerConnector": "AH530P7W",
            "PowerFeedDrop": "ABOVE_RACK",
            "UplinkGbps": "UPLINK_40G",
            "UplinkCount": "UPLINK_COUNT_4",
            "FiberOpticCableType": "MULTI_MODE",
            "OpticalStandard": "OPTIC_40GBASE_SR",
            "MaximumSupportedWeightLbs": "MAX_2000_LBS"
        }
    }
}
```

Make note of the `SiteId` value generated (`os-0431a19da45a53dc0`) – we will use this information later on.

Now that we have a site created, we can create an Outpost. An important distinction to be made is that an *Outpost* is a logical entity that encompasses a certain capacity deployed at a customer site. Capacity is provided by adding *Outposts racks*. An Outpost can have multiple racks providing capacity.

Let's take a look at some screenshots of the Outposts creation process. Click on **Outposts** in the left-hand pane:

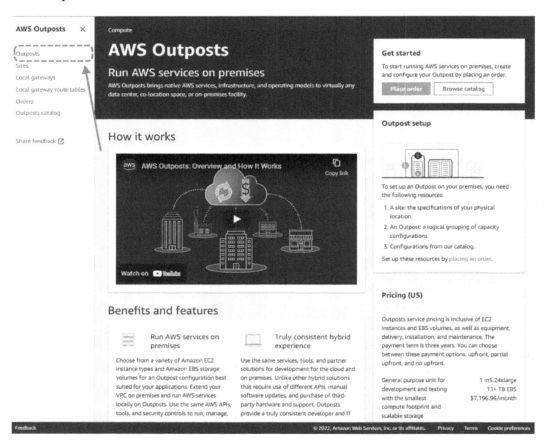

Figure 3.15 – The Outposts menu on the Outposts service landing page

You will be taken to the **Outposts** landing page. Next, click **Create Outpost**:

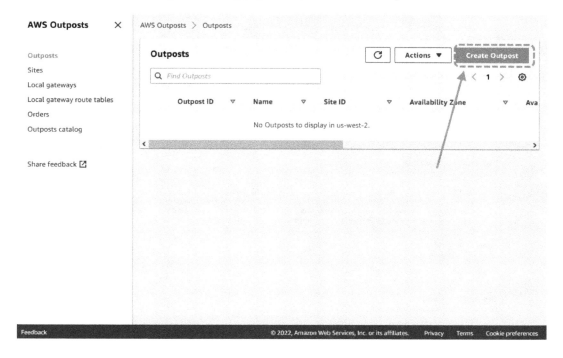

Figure 3.16 – The Outposts list page

At **Create Outpost**, we populate the needed information and click on **Create Outpost**. It is very important to define the AZ and note that all capacity pooled into this Outpost ID will be tied to that single AZ. If your architectural requirement is to consider the failure of an entire AZ, you will need to create another Outpost tied to another AZ in the parent region.

During this step, we will have the option to select the **Site ID** value that we created earlier:

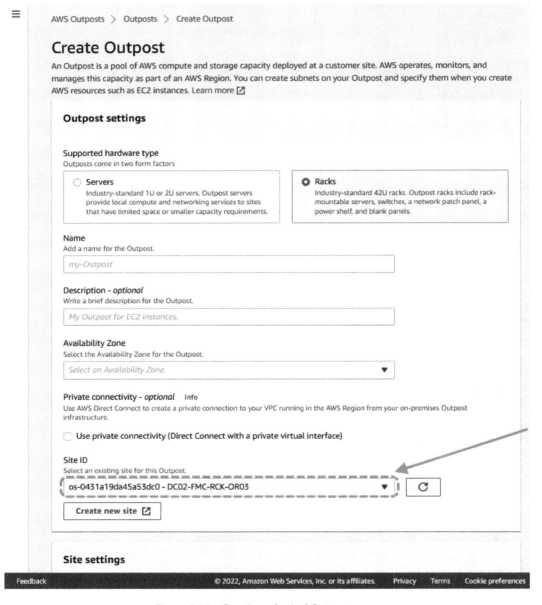

Figure 3.17 – Creating a logical Outpost page

You can populate these fields using the *AWS CLI*. Here, I will provide a sample command.

To create an Outpost, see the following:

```
aws outposts create-outpost \
    --name "DC02-FMC-RCK-OR03-OP1" \
    --description "IT Datacenter 02 in Oregon Outpost 01 AZ03" \
    --site-id os-0431a19da45a53dc0 \
    --availability-zone "us-west-2c" \
    --availability-zone-id "usw2-az3" \
    --tags Name="OR03_DC02_FMT",\
        Status="LAB",\
        CC=`whoami` \
    --supported-hardware-type "RACK";
```

The output will be something similar to what follows:

```
{
    "Outpost": {
        "OutpostId": "op-0fb4c69a3636924b5",
        "OwnerId": "123456789012",
        "OutpostArn": "arn:aws:outposts:us-west-
2:123456789012:outpost/op-0fb4c69a3636924b5",
        "SiteId": "os-0431a19da45a53dc0",
        "Name": "DC02-FMC-RCK-OR03-OP1",
        "Description": "IT Datacenter 02 in Oregon Outpost 01
AZ03",
        "LifeCycleStatus": "PENDING",
        "AvailabilityZone": "us-west-2c",
        "AvailabilityZoneId": "usw2-az3",
        "Tags": {
            "CC": "cloudshell-user",
            "Status": "LAB",
            "Name": "OR03_DC02_FMT"
        },
        "SiteArn": "arn:aws:outposts:us-west-
2:123456789012:site/os-0431a19da45a53dc0",
        "SupportedHardwareType": "RACK"
    }
}
```

This time, note the `OutpostID` value generated (`op-0fb4c69a3636924b5`) – we will use this information later on.

Finally, time to place an order. Let's take a look at screenshots of the process. Click on **Orders** in the left-hand pane:

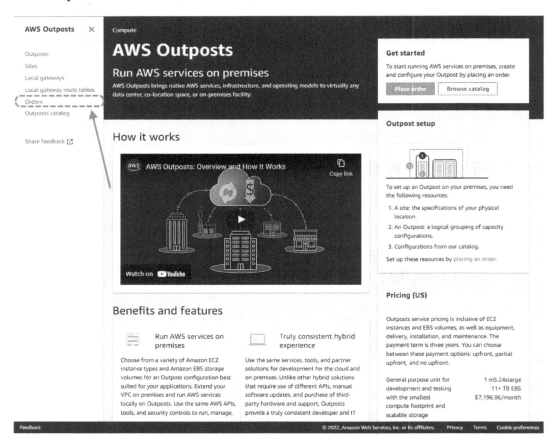

Figure 3.18 – The Orders menu on the Outposts service landing page

You will be taken to the **Orders** landing page. Next, click **Place order**:

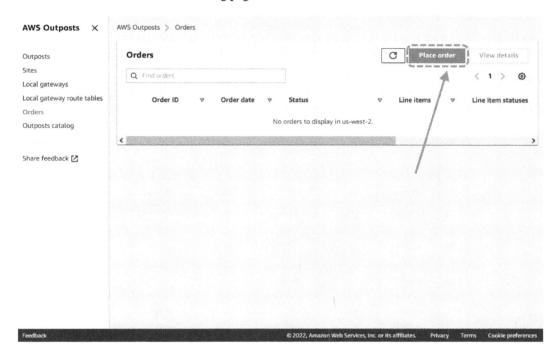

Figure 3.19 – The Orders list page

You will be taken to the **Select configurations** page. Here, you will select the SKU for your order – for the purposes of this walkthrough, let's use a pre-validated SKU. Click **Next**:

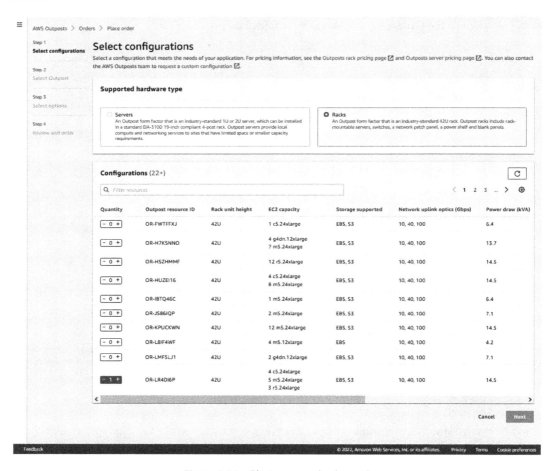

Figure 3.20 – Placing an order (step 1)

In this step, the `Outpost ID` value comes in handy. Select **Use an existing Outpost** and select the desired Outpost:

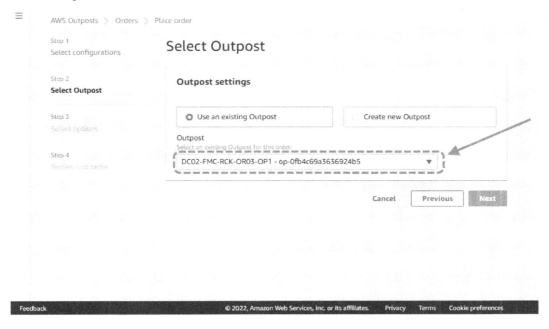

Figure 3.21 – Placing an order (step 2)

Select your **Payment** option and **Shipping address** (it can be different from the **Operating address**). Click **Next**:

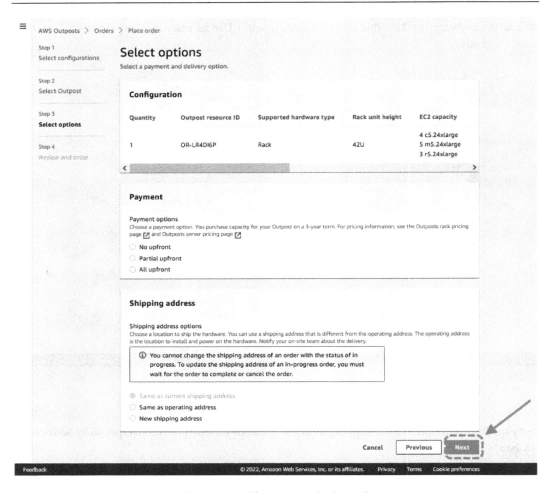

Figure 3.22 – Placing an order (step 3)

Review your selections and click **Place order**:

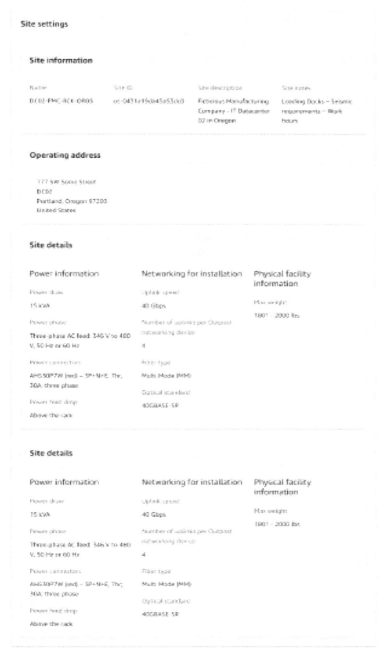

Figure 3.23 – Placing an order (step 4)

The order has now been placed and the clock is ticking to get your order fulfilled. At this time, you must have a thorough understanding of the ordering process, the key concepts involved such as sites, Outposts versus Outpost, and the importance of getting ready in advance because several questions cannot just be responded to by selecting random answers to change later.

If you try to change the information for a site or an Outpost, you can only modify the *name* and *description*. The order creation is dependent on these items and a wrong selection may signify the cancelation of the entire order. In the next chapter, you will discover that ordering Outposts may look reasonably simple with just a handful of questions requiring more work, but in reality, it is a complex and potentially lengthy journey.

Order fulfillment and installation process

With this background, now we should be able to identify whether AWS Outposts is suitable for our edge solution or use case. Are we ready to go over the AWS Console order and AWS Outposts?

Not quite yet – it is easy to assume that we have fulfilled all the prerequisites for the AWS Outposts rack to be shipped to our facility. AWS Outposts will be dispatched as a fully assembled rack in its entirety instead of sending multiple boxes and assembling them onsite. Also, when the rack is delivered to the customer's location, it is not rolled into position and left unattended; in fact, it is accompanied by network technicians who will perform sanity checks. We will talk about this in extensive detail in this section.

Let's begin with this simple fact: suppose that after reading the previous chapters you think it is time to have an Outposts rack in your data center. You read about AWS Outposts – you know what is under the hood. You know AWS technologies and spend quality time on the drawing board to develop an architecture that addresses your use case and is supported by services and capabilities of AWS Outposts. Finally, you headed to the AWS console and identified a pre-validated SKU that meets your needs.

After careful consideration, you checked the AWS pricing page and concluded that AWS Outposts fits the bill. You have several years of data center experience with commodity hardware and you feel you know what it takes to give AWS technicians the needed infrastructure. Nothing is holding you up and you go ahead and order an AWS rack for yourself. Now, the clock is ticking to have it delivered to your facility, right? Almost certainly not.

One cannot simply go to the AWS Console, order a rack, and get it delivered with no further questions asked by the AWS Outposts team, and that applies even if you are an experienced AWS customer and have ordered one before. AWS will not even authorize the manufacturing of the rack until the order is intentional, legitimate, and conscientious.

In accordance with its *customer obsession* leadership principle, AWS will always want to ensure that the customer understands what they are getting into by buying AWS Outposts and what it entails. AWS wants to set its customers up for success and needs to be certain that they are selecting the right tool for the job.

Remember we said how different AWS Outposts is compared to traditional hardware? This is manifested in every aspect of this product. It is not available off-the-shelf and it is not kept in stock ready to be ordered and shipped – not even the pre-validated SKUs are. Every Outpost is a unique order that has to be manufactured after a validation process conducted by the AWS team.

An order placed by mistake is not an option – the chain of events triggered by the order is highly complex, time-consuming, and costly. No one wants to have the dissatisfaction of an Outposts rack being returned, not by a mistakenly placed order, nor by an unhappy customer. That translates to a bad customer experience and this is unacceptable by AWS standards.

The technical validation process consists of a one-on-one engagement between AWS and the customer. This interaction consists of a series of checks and activities whose sole purpose is to ascertain what to expect after ordering an Outpost and that the customer will be ready for what lies ahead by going over the process to get it delivered.

While doing it all yourself is an alternative, AWS strongly prefers to employ its *working-backward* mechanism and directly engage with the customer at the earliest stage possible. AWS envisioned a process to assure smooth selection, order, delivery, and installation of AWS Outposts. Here are the stages.

Pre-sales

At the **pre-sales** stage, AWS technical representatives and the customer account manager organize a meeting with the customer to qualify for the opportunity, a mandatory step whose aim is to assure the following:

- The customer understands the use cases for Outposts and has enough information to decide whether Outposts is a suitable choice for their initiative. That may lead AWS to schedule technical workshops and provide support material to amplify customer knowledge and strengthen their ability to decide with confidence.

- The customer knows AWS technologies. Outposts uses the same concepts, constructs, services, interfaces, and APIs as used in the AWS Region – therefore, having IT personnel with strong knowledge of AWS is a must. If you have a third-party service provider configuring and operating AWS Outposts, it will be their responsibility to provide the staff with AWS skills.

- There will be an assessment of application requirements, boundary conditions, constraints, and potential caveats. Identify the architecture, create a rough estimation of capacity sizing, and identify the installation site. Potentially, a further meeting will be held to dive deeper into these subjects and refine sizing, site requirements, and network design.

Sizing and pricing

In the next step, at the **sizing and pricing** stage, the data and details from the previous stage are used to create a final sizing and to generate a quote that will be submitted to the customer for acceptance. This sizing may fit into an existing AWS Outposts SKU or may be a custom SKU. Remember, with AWS Outposts, you are not ordering physical servers with certain hardware specs – you are ordering *capacity*.

If a custom SKU is required, an internal process is executed to validate the configuration and a specific SKU entry is created and made available for selection using an AWS account indicated by the customer. It's important to mention that this custom SKU will be available for selection only at the account number supplied. For pre-validated SKUs, the customer can order using any account under its control.

Order submission

If the quote is accepted, the customer can log on using AWS Console using the AWS account selected to manage AWS Outposts and perform the **order submission**, which is the next step. One differentiator of this service is that billing only starts when Outposts is delivered and activated. Now, the manufacturing of the rack is authorized and it is time to prepare the site to receive it.

For Outposts servers, this process is simpler, although AWS still suggests you actively engage with the AWS team to ensure you understand what you are buying, your use case is a good match for the product, and what to expect from it. However, there are no custom SKUs and the site requirements are far less restrictive.

Site validation

We now arrive at the **site validation** step. The order is placed and now another AWS team will engage to assist and support the customer to have the site ready and assure a smooth delivery. This process is mandatory and your mileage will vary in terms of complexity.

Some customers take the route to place the rack at a colocation partner site. AWS has a program to develop and authorize colocation partners – you can check with AWS for partners available in your country. If you manage to find one, then great, the process should be significantly expedited. Approved colocation partners have been certified to be able to provide everything an AWS Outposts rack needs in terms of facilities, networking, and power.

If your choice is to bring the rack to your premises, roll up the sleeves and get ready for the exercise. AWS will work with you to make sure you have docks to receive the rack, that it can be rolled to be placed into the final position, and that the facility can provide the power and airflow required. If your country is exposed to seismic events, the rack comes with braces to provide anchorage.

Once these checks are passed, it is time for AWS technicians to schedule an onsite visit for the inspection of the loading dock, walking the path from the loading dock to the rack's final position, inspecting the power and cooling capabilities, and performing networking tests. They will bring their own networking gear and expect to connect to the same fiber termination that will be plugged into the AWS Outposts patch panel once it arrives.

A series of tests will be performed, executing the same steps that will be carried out by the Outposts networking components setup procedure. Establish **BGP** (**Border Gateway Protocol**) sessions, receive routes, bring up VPN tunnels, and connect back to its parent region. The objective is to make sure everything will work without hiccups during the bootstrap process when the rack arrives onsite. AWS will be extremely reluctant to allow the rack to be delivered and stay unassisted, even for one single night.

Order approval

This procedure is expected to occur three to four weeks prior to the customer need-by-date and to be performed in one day. If everything works well, AWS will kick off the **order approval** step where the paperwork and contracts need to be worked out and the mandatory Enterprise Support requirement needs to be arranged and verified. If the parties go all the way through and tick all checkboxes, AWS will propose an installation date aligning with the customer's expectation.

Delivery and installation

The order approval triggers the **delivery and installation** phase. When the day of delivery finally shows in the calendar, two AWS teams will arrive at the designed site, ideally at the earliest hour during the working day. Special conditions and labor outside working hours need to be arranged and negotiated with AWS.

The white glove team walks the path from the loading dock to final placement, once again looking for obstructions or obstacles while ensuring nothing has changed since the site validation. Subsequently, they perform the uncrate of Outposts inside the truck and proceed to maneuver towards the designated position. If bracing was specified, it has to be done by a contractor or customer staff member, but the team will not leave the site until that bracing has been performed.

The infrastructure team will be in charge of feeding power to the rack and connecting network cables. Ideally, the customer should provide an electrician and a network engineer to be available onsite in case of unforeseen events. AWS technicians will never act or intervene in the customer facility elements. Upon completion, the staff leaves the site and performs the handover to the service team to perform the **activation**.

Activation

The service team will act remotely and if the equipment passes all quality and operational tests, it is declared ready for production, and the process is finally completed. The status of the Outposts at the AWS Console will change from **Provisioning** to **Activated**.

From this point on, the billing cycles will start and the order is marked as **Fulfilled**. AWS naturally will not leave the court just because the rack was successfully installed. If the customer needs further assistance, the AWS team will work to ensure the customer is successful in their endeavor.

In this chapter, you navigated through the chain of events and steps needed to successfully have AWS Outposts delivered to you and achieve readiness, be able to orchestrate this process, and understand all its intricacies, dependencies, and pitfalls. It is a long but rewarding journey – your Outposts is activated and ready to rock.

There is an AWS Blog page describing this process, which can be found at this URL: `https://aws.amazon.com/blogs/architecture/field-notes-preparing-for-aws-outposts-ordering-delivery-and-installation/`.

Closing this section, a word on how sensitive this process is – AWS will only authorize the manufacturing of the rack if everything related to this phase is clearly defined, well understood by the customer, and has gone through comprehensive checklists performed together with the customer technical teams.

AWS is bound to its motto of being the Earth's most customer-centric organization and a bad customer experience is not an option. Unless there is a thoroughly defined flight plan for an Outposts order, it will not even leave the hangar.

Summary

In this chapter, we discovered that an Outposts order is not just a matter of going to an online store, selecting a product with some optional components and accessories to be added to a shopping cart, selecting a shipping address and payment method, and finally, clicking on a button to finish the process.

You are now able to identify the pre-validated Outposts SKUs, their requirements and configurations, and what is included in the price. For each AWS service that you can run on Outposts, you know how they are priced and all available options. Moreover, you can estimate data transfer charges based on the configuration you selected.

We have gone through an in-depth walkthrough of the order process at the AWS console, with screenshots and valid entries. AWS CLI command snippets were provided to be used in automated processes or for documentation purposes. Moreover, you can explain the difference between *Outpost* and *Outposts*.

Finally, the chapter gave a comprehensive overview of the installation process. If an Outposts order is approved, the installation process kicks in and it is far from simple: you have to prove to AWS that you are ready to receive your order.

Now that you are empowered to have AWS Outposts delivered to you, it's finally time to get your hands dirty and configure your Outposts to be ready for action and run AWS services in the next chapter.

Part 2:
Security,
Monitoring, and
Maintenance

This part covers three very important areas to master when your Outposts is finally handed over to the customer. You will get hands-on with Outposts and take it for a spin to see how it works in real life, discover the built-in security features, and see how the product uses AWS Security services in the region, concluding with how to monitor your Outposts to ensure business continuity, maximizing the value delivered by the product.

This part has the following chapters:

- *Chapter 4, Operations and Working with Outposts Resources*
- *Chapter 5, Security Aspects in Outposts*
- *Chapter 6, Monitoring Outposts*

4

Operations and Working with Outposts Resources

You finally got your AWS Outposts delivered to your data center, installed, provisioned, and activated. The AWS Outposts team performed the handover process and you are hopefully excited to see it in action.

We have a lot to cover in this chapter and it will be heavily hands-on. The AWS Management Console is the interface of choice when we are beginning with AWS, but as we progress in our usage of the platform, there is a paradigm shift in the way we configure and implement changes to prefer **Infrastructure as Code (IaC)** tools such as Terraform and CloudFormation.

These tools alone are worthy of a book dedicated to teaching how to use them, falling out of the scope of this book. However, the AWS CLI is simpler and more familiar, because it uses the same scripting principles and techniques consecrated by the likes of Linux Bash or Microsoft PowerShell.

Once again, you don't need to install the AWS CLI on your computer. We will use AWS CloudShell, a ready-to-use, easily provisioned environment to perform our API calls. In this chapter, you will learn the following:

- How to perform the initial setup of your Outposts
- How to design a network infrastructure to support complex scenarios
- How to launch and log in to an Outposts EC2 instance
- How to operate and configure the local gateway
- How to share Outposts resources with other accounts

Working with instances

It has been an amazing journey so far. We have a thorough understanding of Outposts, the anatomy of the rack, and how to select, order, and have one delivered to us. It is time to head on to the AWS

Console and play with our Outposts, which should now show the status as **Active** in the service page summary:

Figure 4.1 – AWS Outposts Summary page

Before we move further, there is a very important distinction to make: *Availability Zone* versus *Availability Zone ID*. As we know, AWS creates an abstraction to prevent Availability Zones from being saturated by mapping real Availability Zone IDs (*usw2-az1*, shown in *Figure 4.1*) to different Availability Zones in the console (*us-west-2a*, shown in *Figure 4.1*) for distinct AWS accounts.

As an example, you can have two AWS accounts, *A* and *B*. AWS account *A* can have the mapping mentioned earlier, and AWS account *B* can have the *us-west-2a* Availability Zone mapped to Availability Zone ID *usw2-az3*, for example. This allows for better distribution across the various Availability Zones, as there is a natural tendency to begin building architectures with *Zone 01*, which could lead to excessive use of resources while *Zone 03*, for example, is underused.

This is going to be a heavy, hands-on chapter, so buckle up! As professionals, we always must be cognizant of best practices. For AWS, the best practices are compiled in the AWS Well-Architected Framework, which can be found in several languages at this URL: `https://wa.aws.amazon.com/`.

The *general design principles* of the Well-Architected Framework are the mandates of good design and secure, scalable, resilient, and high-performing architectures in the cloud. As highlighted by the principle *Automate to make architectural experimentation easier*, you can find these principles in the English language at this URL `https://wa.aws.amazon.com/wat.design_principles.wa-dp.en.html`. We will make extensive use of AWS CLI for our configurations throughout this chapter. Once again, we will leverage AWS CloudShell to run commands.

To configure Outposts, the key parameter is `OutpostId`, which was generated when we created our Outpost. Let's remember, an *Outpost* is a logical entity that encompasses a certain capacity deployed at a customer site. How do we discover `OutpostId` using the CLI? Our test environment for this session was deployed in the Oregon region, `us-west-2`.

List existing AWS Outposts and their respective IDs for the Oregon region:

```
aws outposts list-outposts --region us-west-2
```

The output will be something like this:

```
{
    "Outposts": [
        {
            "OutpostId": "op-8b286039iedad23e0",
            "OwnerId": "123456789012",
            "OutpostArn": "arn:aws:outposts:us-west-
2:123456789012:outpost/op-8b286039iedad23e0",
            "SiteId": "os-2be553a550d645940",
            "Name": "DC02-FMC-RCK-OR01-OP2",
            "Description": "IT Datacenter 02 in Oregon Outpost
02 AZ01",
            "LifeCycleStatus": "ACTIVE",
            "AvailabilityZone": "us-west-2a",
            "AvailabilityZoneId": "usw2-az1",
            "Tags": {},
            "SiteArn": "arn:aws:outposts:us-west-
2:123456789012:site/os-2be553a550d645940",
            "SupportedHardwareType": "RACK"
        }
    ]
}
```

Let's make note of our Outpost ID, op-8b286039iedad23e0, and the Availability Zone ID where the rack is anchored, which in our case is usw2-az1. Subsequently, we would like to know what instance types our Outposts are capable of running, and here is how to find out. To create a subnet associated with AWS Outposts, you will need the Outpost **Amazon Resource Name** (**ARN**): arn:aws:outposts:us-west-2:123456789012:outpost/op-8b286039iedad23e0.

List instance types available for an Outpost ID:

```
aws outposts get-outpost-instance-types --outpost-id
op-8b286039iedad23e0
```

The output will be something like this:

```
{
    "InstanceTypes": [
        {
            "InstanceType": "c5.4xlarge"
        },{
            "InstanceType": "c5.large"
        },{
            "InstanceType": "g4dn.2xlarge"
        },{
            "InstanceType": "i3en.3xlarge"
        },{
            "InstanceType": "m5.2xlarge"
        },{
            "InstanceType": "m5.4xlarge"
        },{
            "InstanceType": "m5.xlarge"
        },{
            "InstanceType": "m5d.12xlarge"
        },{
            "InstanceType": "r5.24xlarge"
        },{
            "InstanceType": "r5.4xlarge"
        }
    ],
    "OutpostId": "op-8b286039iedad23e0",
    "OutpostArn": "arn:aws:outposts:us-west-
2:123456789012:outpost/op-8b286039iedad23e0"
}
```

Let's select one instance type available to launch one test instance on our Outpost ID, op-8b286039iedad23e0; for the purposes of our lab, let's choose m5d.12xlarge. The specifications for this instance type are 48 vCPUs, 192 GiB memory, 2 x 900 NVMe SSD, up to 12 Gbps network performance, and up to 9.500 Gbps EBS bandwidth.

Outposts networking design example

After activation, there are some steps to be taken for your Outposts to be fully functional. Every Outposts rack will always ship with EC2 and EBS capabilities. To leverage these services, the network design needs to be implemented, and that's exactly what we will do next.

For our lab, we will use two distinct **Virtual Private Clouds** (**VPCs**). Each VPC will contain two **Classless Inter-Domain Routing** (**CIDR**) blocks, and each CIDR block will contain four subnets. Two subnets will map to the Outposts **Availability Zone** (**AZ**), which in our case is usw2-az1, and the other two will map to distinct AZs.

The reason for this architecture is to demonstrate that in the AZ where Outposts is anchored, we can create subnets that exist inside the Outposts rack and subnets that exist only in the Region. This is the critical concept to understand how the AWS control plane knows that a certain resource (EC2, RDS, and so on) must be launched inside Outposts. When you indicate to launch the resource using a subnet that is mapped to an Outpost ID, it will be launched inside the rack or server.

The following is the networking design we are going to implement. Keep this information handy as you go through the chapter; it will be used in several commands, and you can always refer back to these parameters to understand what is being done in each AWS CLI command:

Outpost ID: op-8b286039iedad23e0

Outposts anchor AZ: usw2-az1

Outpost ARN: arn:aws:outposts:us-west-2:123456789012:outpost/op-8b286039iedad23e0

VPC [01]:

- CIDR block [1]: 10.20.0.0/21:

  - Subnet [01] AZ ID [usw2-az1]: 10.20.0.0/23
  - Subnet [02] AZ ID [usw2-az2]: 10.20.2.0/23
  - Subnet [03] AZ ID [usw2-az3]: 10.20.4.0/23
  - Subnet [04] AZ ID [usw2-az1]: 10.20.6.0/23 » op-8b286039iedad23e0

- CIDR block [2]: 10.20.8.0/21:

  - Subnet [01] AZ ID [usw2-az1]: 10.20.8.0/23
  - Subnet [02] AZ ID [usw2-az2]: 10.20.10.0/2
  - Subnet [03] AZ ID [usw2-az3]: 10.20.12.0/2
  - Subnet [04] AZ ID [usw2-az1]: 10.20.14.0/23 » op-8b286039iedad23e0

VPC [02]:

- CIDR block [1]: `10.20.40.0/21`:

  - Subnet [01] AZ ID [usw2-az1]: `10.20.40.0/23`

  - Subnet [02] AZ ID [usw2-az2]: `10.20.42.0/23`

  - Subnet [03] AZ ID [usw2-az3]: `10.20.44.0/23`

  - Subnet [04] AZ ID [usw2-az1]: `10.20.46.0/23` » `op-8b286039iedad23e0`

- CIDR block [2]: `10.20.48.0/21`:

  - Subnet [01] AZ ID [usw2-az1]: `10.20.48.0/23`

  - Subnet [02] AZ ID [usw2-az2]: `10.20.50.0/23`

  - Subnet [03] AZ ID [usw2-az3]: `10.20.52.0/23`

  - Subnet [04] AZ ID [usw2-az1]: `10.20.54.0/23` » `op-8b286039iedad23e0`

Let's create VPC [01] using AWS CLI. The first thing to be created, if you want your VPC to communicate with the internet using public or elastic IP addresses, is to create an **Internet Gateway** (**IGW**); more details can be found at this URL: `https://docs.aws.amazon.com/vpc/latest/userguide/VPC_Internet_Gateway.html`.

Create an IGW:

```
aws ec2 create-internet-gateway \
    --query "InternetGateway.InternetGatewayId" \
    --output text \
    --region "us-west-2";
```

The output will be something like this:

```
igw-0301e72c4522f4e2e
```

Now, create the VPC by running the following command:

```
aws ec2 create-vpc \
    --cidr-block 10.20.0.0/21 \
    --instance-tenancy default \
    --no-amazon-provided-ipv6-cidr-block \
    --query "Vpc.
[VpcId,DhcpOptionsId,CidrBlockAssociationSet[*].AssociationId]"
\
```

```
    --output text \
    --region "us-west-2";
```

The output will be something like this, respectively, as defined in the query statement: *VPC ID*, *DHCP Options ID*, and *VPC CIDR Association ID*:

```
vpc-06b64f761be00fcb3      dopt-04dfb9f1946611d1c      vpc-cidr-
assoc-06dece94e16fe573e
```

With these constructs ready, we can attach the IGW with the VPC using the following command:

```
aws ec2 attach-internet-gateway \
    --internet-gateway-id igw-0301e72c4522f4e2e \
    --vpc-id vpc-06b64f761be00fcb3 \
    --region "us-west-2";
```

We can check the association in the AWS Management Console:

Figure 4.2 – AWS Console VPC page – Internet Gateways

Moving further, let's create our subnets. To describe all the AZs that can be used in a given AWS Region, use this command:

```
aws ec2 describe-availability-zones \
    --query "AvailabilityZones[].{ZoneNameId:ZoneId}" \
    --output text \
```

```
    --region "us-west-2";
```

The output will be something like this:

```
usw2-az1
usw2-az2
usw2-az3
usw2-az4
```

To create subnet [01] for VPC [01], whose VPC ID is vpc-06b64f761be00fcb3 associated with Availability Zone ID usw2-az1 using CIDR block [1], this is the command:

```
aws ec2 create-subnet \
    --vpc-id "vpc-06b64f761be00fcb3" \
    --availability-zone-id "usw2-az1" \
    --cidr-block 10.20.0.0/23 \
    --region "us-west-2";
```

The output will be something like this:

```
{
    "Subnet": {
        "AvailabilityZone": "us-west-2a",
        "AvailabilityZoneId": "usw2-az1",
        "AvailableIpAddressCount": 507,
        "CidrBlock": "10.20.0.0/23",
        "DefaultForAz": false,
        "MapPublicIpOnLaunch": false,
        "State": "available",
        "SubnetId": "subnet-0603cf420b7e30598",
        "VpcId": "vpc-06b64f761be00fcb3",
        "OwnerId": "123456789012",
        "AssignIpv6AddressOnCreation": false,
        "Ipv6CidrBlockAssociationSet": [],
        "SubnetArn": "arn:aws:ec2:us-west-
2:123456789012:subnet/subnet-0603cf420b7e30598",
        "EnableDns64": false,
        "Ipv6Native": false,
        "PrivateDnsNameOptionsOnLaunch": {
```

```
            "HostnameType": "ip-name",
            "EnableResourceNameDnsARecord": false,
            "EnableResourceNameDnsAAAARecord": false
        }
    }
}
```

Make note of the newly created subnet ID, `subnet-0603cf420b7e30598`, and corresponding Availability Zone ID, `usw2-az1`. To create the remaining two subnets for this CIDR block, repeat the command adjusting the `--availability-zone-id` parameter accordingly, and also make note of the new subnet IDs. In my case, after running the commands, these were the newly created subnets:

- `10.20.2.0/23 - subnet-0aa6f2a902b6fbfee - usw2-az2`
- `10.20.4.0/23 - subnet-098667c9f03b73f69 - usw2-az3`

Now, to create the subnet that must be associated with AWS Outposts in CIDR block [1], this is the command:

```
aws ec2 create-subnet \
    --vpc-id "vpc-06b64f761be00fcb3" \
    --availability-zone-id "usw2-az1" \
    --outpost-arn "arn:aws:outposts:us-west-
2:123456789012:outpost/op-8b286039iedad23e0" \
    --cidr-block 10.20.6.0/23 \
    --region "us-west-2";
```

The output will be something like this:

```
{
    "Subnet": {
        "AvailabilityZone": "us-west-2a",
        "AvailabilityZoneId": "usw2-az1",
        "AvailableIpAddressCount": 507,
        "CidrBlock": "10.20.6.0/23",
        "DefaultForAz": false,
        "MapPublicIpOnLaunch": false,
        "State": "available",
        "SubnetId": "subnet-061fe41c63e699bb9",
        "VpcId": "vpc-06b64f761be00fcb3",
```

```
        "OwnerId": "123456789012",
        "AssignIpv6AddressOnCreation": false,
        "Ipv6CidrBlockAssociationSet": [],
        "SubnetArn": "arn:aws:ec2:us-west-
2:123456789012:subnet/subnet-061fe41c63e699bb9",
        "OutpostArn": "arn:aws:outposts:us-west-
2:123456789012:outpost/op-8b286039iedad23e0",
        "EnableDns64": false,
        "Ipv6Native": false,
        "PrivateDnsNameOptionsOnLaunch": {
            "HostnameType": "ip-name",
            "EnableResourceNameDnsARecord": false,
            "EnableResourceNameDnsAAAARecord": false
        }
    }
}
```

Before we create the subnets using CIDR block [2], we need to associate this block with VPC [01] using this command:

```
aws ec2 associate-vpc-cidr-block \
    --vpc-id "vpc-06b64f761be00fcb3" \
    --cidr-block 10.20.8.0/21 \
    --region "us-west-2";
```

The output will be something like this:

```
{
    "CidrBlockAssociation": {
        "AssociationId": "vpc-cidr-assoc-0148b71b0d4ccae0d",
        "CidrBlock": "10.20.8.0/21",
        "CidrBlockState": {
            "State": "associating"
        }
    },
    "VpcId": "vpc-06b64f761be00fcb3"
}
```

To create subnet [01] for VPC [01], whose VPC ID is vpc-06b64f761be00fcb3 associated with Availability Zone ID usw2-az1 using CIDR block [2], this is the command:

```
aws ec2 create-subnet \
    --vpc-id "vpc-06b64f761be00fcb3" \
    --availability-zone-id " usw2-az1" \
    --cidr-block 10.20.8.0/23 \
    --region "us-west-2";
```

The output will be something like this:

```
{
    "Subnet": {
        "AvailabilityZone": "us-west-2a",
        "AvailabilityZoneId": "usw2-az1",
        "AvailableIpAddressCount": 507,
        "CidrBlock": "10.20.8.0/23",
        "DefaultForAz": false,
        "MapPublicIpOnLaunch": false,
        "State": "available",
        "SubnetId": "subnet-068754e6ff530edd4",
        "VpcId": "vpc-06b64f761be00fcb3",
        "OwnerId": "123456789012",
        "AssignIpv6AddressOnCreation": false,
        "Ipv6CidrBlockAssociationSet": [],
        "SubnetArn": "arn:aws:ec2:us-west-
2:123456789012:subnet/subnet-068754e6ff530edd4",
        "EnableDns64": false,
        "Ipv6Native": false,
        "PrivateDnsNameOptionsOnLaunch": {
            "HostnameType": "ip-name",
            "EnableResourceNameDnsARecord": false,
            "EnableResourceNameDnsAAAARecord": false
        }
    }
}
```

Make note of the newly created subnet ID, subnet-068754e6ff530edd4, and corresponding Availability Zone ID, usw2-az1. To create the remaining two subnets for this CIDR block, repeat the command adjusting the --availability-zone-id parameter accordingly, and also make note of the new subnet IDs. In my case, after running the commands, these were the newly created subnets:

- 10.20.10.0/23 - subnet-0a00bf357a5d4ef7f - usw2-az2
- 10.20.12.0/23 - subnet-0425aca6373595062 - usw2-az3

Now, to create the subnet that must be associated with AWS Outposts in CIDR block [2], this is the command:

```
aws ec2 create-subnet \
    --vpc-id "vpc-06b64f761be00fcb3" \
    --availability-zone-id "usw2-az1" \
    --cidr-block 10.20.14.0/23 \
    --outpost-arn "arn:aws:outposts:us-west-
2:123456789012:outpost/op-8b286039iedad23e0" \
    --region "us-west-2";
```

The output will be something like this:

```
{
    "Subnet": {
        "AvailabilityZone": "us-west-2a",
        "AvailabilityZoneId": "usw2-az1",
        "AvailableIpAddressCount": 507,
        "CidrBlock": "10.20.14.0/23",
        "DefaultForAz": false,
        "MapPublicIpOnLaunch": false,
        "State": "available",
        "SubnetId": "subnet-04cc4a56110d117e4",
        "VpcId": "vpc-06b64f761be00fcb3",
        "OwnerId": "123456789012",
        "AssignIpv6AddressOnCreation": false,
        "Ipv6CidrBlockAssociationSet": [],
        "SubnetArn": "arn:aws:ec2:us-west-
2:123456789012:subnet/subnet-04cc4a56110d117e4",
        "OutpostArn": "arn:aws:outposts:us-west-
2:123456789012:outpost/op-8b286039iedad23e0",
```

```
    "EnableDns64": false,
    "Ipv6Native": false,
    "PrivateDnsNameOptionsOnLaunch": {
        "HostnameType": "ip-name",
        "EnableResourceNameDnsARecord": false,
        "EnableResourceNameDnsAAAARecord": false
    }
  }
}
```

The process to create the second VPC is the same, and you will end up with a new IGW ID, new VPC ID, and new subnet IDs. Here is my VPC table after going through the process:

Figure 4.3 – AWS Console VPC page – Internet Gateways

And this is my final subnet table, after going through all the processes described previously:

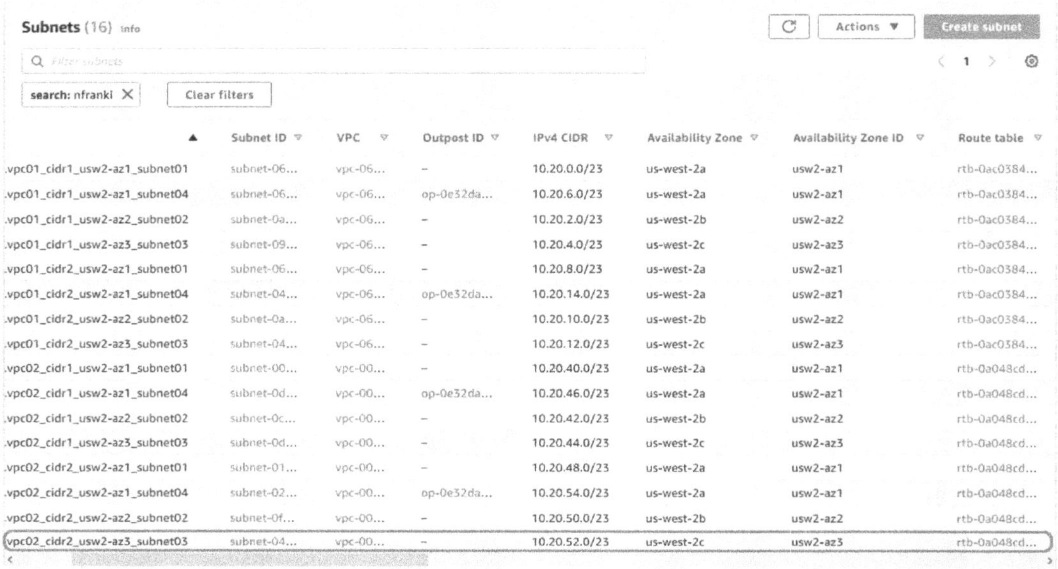

		Subnet ID ▽	VPC ▽	Outpost ID ▽	IPv4 CIDR ▽	Availability Zone ▽	Availability Zone ID ▽	Route table ▽
.vpc01_cidr1_usw2-az1_subnet01	▲	subnet-06...	vpc-06...	–	10.20.0.0/23	us-west-2a	usw2-az1	rtb-0ac0384...
.vpc01_cidr1_usw2-az1_subnet04		subnet-06...	vpc-06...	op-0e32da...	10.20.6.0/23	us-west-2a	usw2-az1	rtb-0ac0384...
.vpc01_cidr1_usw2-az2_subnet02		subnet-0a...	vpc-06...	–	10.20.2.0/23	us-west-2b	usw2-az2	rtb-0ac0384...
.vpc01_cidr1_usw2-az3_subnet03		subnet-09...	vpc-06...	–	10.20.4.0/23	us-west-2c	usw2-az3	rtb-0ac0384...
.vpc01_cidr2_usw2-az1_subnet01		subnet-06...	vpc-06...	–	10.20.8.0/23	us-west-2a	usw2-az1	rtb-0ac0384...
.vpc01_cidr2_usw2-az1_subnet04		subnet-04...	vpc-06...	op-0e32da...	10.20.14.0/23	us-west-2a	usw2-az1	rtb-0ac0384...
.vpc01_cidr2_usw2-az2_subnet02		subnet-0a...	vpc-06...	–	10.20.10.0/23	us-west-2b	usw2-az2	rtb-0ac0384...
.vpc01_cidr2_usw2-az3_subnet03		subnet-04...	vpc-06...	–	10.20.12.0/23	us-west-2c	usw2-az3	rtb-0ac0384...
.vpc02_cidr1_usw2-az1_subnet01		subnet-00...	vpc-00...	–	10.20.40.0/23	us-west-2a	usw2-az1	rtb-0a048cd...
.vpc02_cidr1_usw2-az1_subnet04		subnet-0d...	vpc-00...	op-0e32da...	10.20.46.0/23	us-west-2a	usw2-az1	rtb-0a048cd...
.vpc02_cidr1_usw2-az2_subnet02		subnet-0c...	vpc-00...	–	10.20.42.0/23	us-west-2b	usw2-az2	rtb-0a048cd...
.vpc02_cidr1_usw2-az3_subnet03		subnet-0d...	vpc-00...	–	10.20.44.0/23	us-west-2c	usw2-az3	rtb-0a048cd...
.vpc02_cidr2_usw2-az1_subnet01		subnet-01...	vpc-00...	–	10.20.48.0/23	us-west-2a	usw2-az1	rtb-0a048cd...
.vpc02_cidr2_usw2-az1_subnet04		subnet-02...	vpc-00...	op-0e32da...	10.20.54.0/23	us-west-2a	usw2-az1	rtb-0a048cd...
.vpc02_cidr2_usw2-az2_subnet02		subnet-0f...	vpc-00...	–	10.20.50.0/23	us-west-2b	usw2-az2	rtb-0a048cd...
vpc02_cidr2_usw2-az3_subnet03		subnet-04...	vpc-00...	–	10.20.52.0/23	us-west-2c	usw2-az3	rtb-0a048cd...

Figure 4.4 – AWS Console VPC page – Subnets

Time to have to take a peek at what we are going to do. We will launch one EC2 instance inside the Outpost. This instance will have one private IPv4 and one public IPv4 address assigned automatically. We will log into the instance shell using SSH originating a connection from our AWS CloudShell session targeting the public IPv4 instance address.

The instance will be launched associated with subnet `vpc02_cidr2_usw2-az1_subnet04` highlighted in *Figure 4.4*. The public IP address will be randomly assigned by AWS using one address from its public IP ranges; more information can be found at this URL: `https://docs.aws. amazon.com/general/latest/gr/aws-ip-ranges.html`.

Let's take a peek at the scenario. So far, we configured the network substrate and the IGWs, VPCs, CIDR blocks, and multiple subnets. Each subnet is always associated with just one AZ defined at creation time and it cannot be changed later. The following is a graphical view of what we did up to this point:

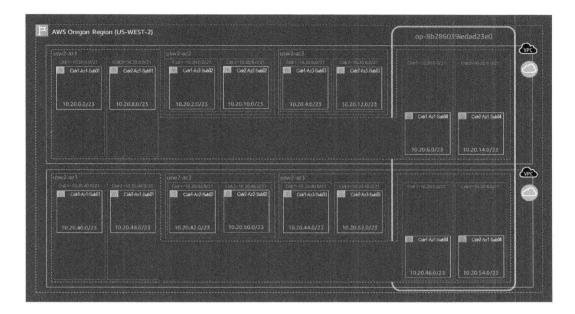

Figure 4.5 – Network design to be implemented in this section

Note that four subnets exist inside Outposts, and we will make them publicly accessible by adding a route entry on the route table pointing to the IGW attached to the VPC. You can create multiple route tables, and one route table can be associated with multiple subnets. However, a subnet can be associated with just one route table and this can be changed later.

One important detail of this lab is that we will use the default route table created whenever we create a VPC. The consequence is that this route table will be associated with any newly created subnet by default. Best practices for security dictate that the default route table should be the most restricted possible, only allowing intra-VPC network traffic.

In the next steps, we will associate the default route (0.0.0.0/0) with the IGW ID in the default route table. Because all subnets are associated with this route table, you can create resources that have incoming and outgoing access to the internet. All they need is to have a public IPv4 address associated. This is not the ideal security practice but it will simplify the lab; in production, you should create custom route tables for public subnets.

We will configure only the subnets that map to our Outposts to automatically assign public IPv4 addresses. These IP addresses can be manually assigned, removed, or re-assigned after the EC2 instance launch. This is a depiction of our scenario:

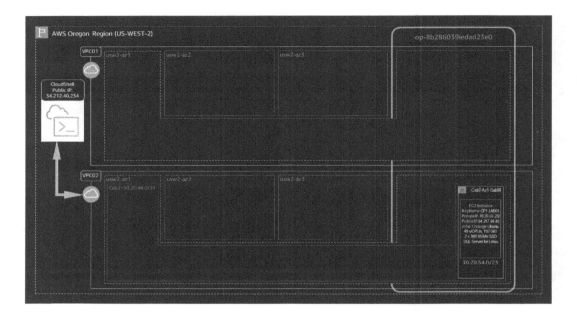

Figure 4.6 – Lab scenario to be implemented in this section

Now, we need to configure our Outposts subnets to auto-assign public IPv4 addresses. We will connect to our instance running inside AWS Outposts performing an SSH to a public IPv4 IP address. For this purpose, we will modify four subnets that live inside Outposts, as shown in *Figure 4.5* with a blue padlock icon, identified by the following IDs:

- subnet-061fe41c63e699bb9 (10.20.6.0/23)
- subnet-04cc4a56110d117e4 (10.20.14.0/23)
- subnet-0dad72c96085078e5 (10.20.46.0/23)
- subnet-02a96f65e7a82ee7c (10.20.54.0/23)

Here is the command executed for the first subnet:

```
aws ec2 modify-subnet-attribute --subnet-id subnet-
061fe41c63e699bb9 \
    --map-public-ip-on-launch \
    --region "us-west-2";
```

There is no output, but we can check with another command:

```
aws ec2 describe-subnets --subnet-ids subnet-061fe41c63e699bb9
\
```

```
    --query "Subnets[].
{AutomaticPublicIPV4Assignment:MapPublicIpOnLaunch}" \
    --region "us-west-2";
```

The output will be something like this:

```
[
    {
        "AutomaticPublicIPV4Assignment": true
    }
]
```

Repeat the process for all other subnets modifying the --subnet-ids parameter. All four subnets associated with our Outpost ID will provide public IPv4 addresses to newly launched instances by default.

Let's move on! When a VPC is created, by default one route table is also created and it is automatically associated with any new subnet. This route table contains the implicit routing that allows all subnets created in a given VPC to mutually communicate by default.

How do we find the IDs for these constructs? You guessed, using AWS CLI, and we need the VPC IDs; in my case, they are vpc-06b64f761be00fcb3 and vpc-00bc84d9f66dd56f0. Here is the command:

```
aws ec2 describe-route-tables \
    --filters "Name=vpc-id,Values=vpc-06b64f761be00fcb3,vpc-
00bc84d9f66dd56f0" \
    --region "us-west-2";
```

The output will be something like this:

```
{
    "RouteTables": [
        {
            "Associations": [
                {
                    "Main": true,
                    "RouteTableAssociationId": "rtbassoc-
0c1d9e809ee9d28fc",
                    "RouteTableId": "rtb-0a048cd1f56c318f6",
                    "AssociationState": {
                        "State": "associated"
```

```
                }
            }
        ],
        "PropagatingVgws": [],
        "RouteTableId": "rtb-0a048cd1f56c318f6",
        "Routes": [
            {
                "DestinationCidrBlock": "10.20.40.0/21",
                "GatewayId": "local",
                "Origin": "CreateRouteTable",
                "State": "active"
            },{
                "DestinationCidrBlock": "10.20.48.0/21",
                "GatewayId": "local",
                "Origin": "CreateRouteTable",
                "State": "active"
            }
        ],
        "Tags": [],
        "VpcId": "vpc-00bc84d9f66dd56f0",
        "OwnerId": "123456789012"
    },{
        "Associations": [
            {
                "Main": true,
                "RouteTableAssociationId": "rtbassoc-
021beb10ead84ae0f",
                "RouteTableId": "rtb-0ac038498f8affc9b",
                "AssociationState": {
                    "State": "associated"
                }
            }
        ],
        "PropagatingVgws": [],
        "RouteTableId": "rtb-0ac038498f8affc9b",
        "Routes": [
```

```
        {
                "DestinationCidrBlock": "10.20.0.0/21",
                "GatewayId": "local",
                "Origin": "CreateRouteTable",
                "State": "active"
        },{
                "DestinationCidrBlock": "10.20.8.0/21",
                "GatewayId": "local",
                "Origin": "CreateRouteTable",
                "State": "active"
        }
    ],
    "Tags": [],
    "VpcId": "vpc-06b64f761be00fcb3",
    "OwnerId": "123456789012"
        }
    ]
}
```

This is the full output for reference. Note the DestinationCidrBlock parameter, which indicates the CIDR blocks associated with a given VPC, and GatewayId, which indicates local (the implicit routing). We are interested in the RouteTable IDs, so let's filter them out:

```
aws ec2 describe-route-tables \
    --filters "Name=vpc-id,Values=vpc-06b64f761be00fcb3,vpc-
00bc84d9f66dd56f0" \
    --query "RouteTables[].
{IdOfTheRouteTable:RouteTableId,IdOfTheVPC:VpcId}" \
    --region "us-west-2";
```

The output will be something like this:

```
[
    {
        "IdOfTheRouteTable": "rtb-0a048cd1f56c318f6",
        "IdOfTheVPC": "vpc-00bc84d9f66dd56f0"
    },{
        "IdOfTheRouteTable": "rtb-0ac038498f8affc9b",
        "IdOfTheVPC": "vpc-06b64f761be00fcb3"
```

```
        }
    ]
```

We need to associate these two route tables with an IGW. By virtue of this association, bidirectional internet traffic will be enabled for these two route tables. Once again, how do we find IGW IDs?

You may have been precautious, perhaps you noted them down somewhere for your reference because we created these IGWs early in the chapter. If you did that, congratulations, as this is good practice. However, the AWS CLI might be able to help once more. Here is the command:

```
aws ec2 describe-internet-gateways \
    --filters "Name=attachment.vpc-id,Values=vpc-
00bc84d9f66dd56f0,vpc-06b64f761be00fcb3" \
    --query "InternetGateways[].
{IdOfTheInternetGateway:InternetGatewayId,IdOfTheVPC:Attachment
s[].VpcId}" \
    --region "us-west-2";
```

The output will be something like this:

```
    [
        {
            "IdOfTheInternetGateway": "igw-0301e72c4522f4e2e",
            "IdOfTheVPC": [
                "vpc-06b64f761be00fcb3"
            ]
        },{
            "IdOfTheInternetGateway": "igw-0677dfe46be4c31d0",
            "IdOfTheVPC": [
                "vpc-00bc84d9f66dd56f0"
            ]
        }
    ]
```

For the next step, we need to associate the quad zero route (0.0.0.0/0) pointing to the IGW ID on our route tables to be able to send and receive a packet from the internet. Here is how:

```
aws ec2 create-route --destination-cidr-block "0.0.0.0/0" \
    --gateway-id igw-0301e72c4522f4e2e \
    --route-table-id rtb-0ac038498f8affc9b \
    --region "us-west-2";
```

The output is very simple:

```
{
    "Return": true
}
```

Just repeat the command, changing the `--gateway-id` and `--route-table-id` parameters accordingly. We are almost there, the **network access control list (NACL)** created by default does not restrict traffic, therefore we should be good to go with nothing being blocked. Our last step is to adjust the security groups. You probably guessed that there is one more `describe` command to execute; let's jump into it:

```
aws ec2 describe-security-groups \
    --filters "Name=vpc-id,Values=vpc-00bc84d9f66dd56f0,vpc-
06b64f761be00fcb3" \
    --region "us-west-2";
```

The full output will be something like this:

```
{
    "SecurityGroups": [
        {
            "Description": "default VPC security group",
            "GroupName": "default",
            "IpPermissions": [
                {
                    "IpProtocol": "-1",
                    "IpRanges": [],
                    "Ipv6Ranges": [],
                    "PrefixListIds": [],
                    "UserIdGroupPairs": [
                        {
                            "GroupId": "sg-049e93c5df3e056dc",
                            "UserId": "123456789012"
                        }
                    ]
                }
            ],
            "OwnerId": "123456789012",
```

```
                    "GroupId": "sg-049e93c5df3e056dc",
                    "IpPermissionsEgress": [
                        {
                            "IpProtocol": "-1",
                            "IpRanges": [
                                {
                                    "CidrIp": "0.0.0.0/0"
                                }
                            ],
                            "Ipv6Ranges": [],
                            "PrefixListIds": [],
                            "UserIdGroupPairs": []
                        }
                    ],
                    "Tags": [
                    ],
                    "VpcId": "vpc-00bc84d9f66dd56f0"
            },{
                    "Description": "default VPC security group",
                    "GroupName": "default",
                    "IpPermissions": [
                        {
                            "IpProtocol": "-1",
                            "IpRanges": [],
                            "Ipv6Ranges": [],
                            "PrefixListIds": [],
                            "UserIdGroupPairs": [
                                {
                                    "GroupId": "sg-09cfc574cf19a85cc",
                                    "UserId": "123456789012"
                                }
                            ]
                        }
                    ],
                    "OwnerId": "123456789012",
                    "GroupId": "sg-09cfc574cf19a85cc",
```

```
            "IpPermissionsEgress": [
                {
                    "IpProtocol": "-1",
                    "IpRanges": [
                        {
                            "CidrIp": "0.0.0.0/0"
                        }
                    ],
                    "Ipv6Ranges": [],
                    "PrefixListIds": [],
                    "UserIdGroupPairs": []
                }
            ],
            "Tags": [
            ],
            "VpcId": "vpc-06b64f761be00fcb3"
        }
    ]
}
```

Here is the command to filter out the security group IDs:

```
aws ec2 describe-security-groups \
    --filters "Name=vpc-id,Values=vpc-00bc84d9f66dd56f0,vpc-
06b64f761be00fcb3" \
    --query "SecurityGroups[].
{IdOfTheSecurityGroup:GroupId,IdOfTheVPC:VpcId}" \
    --region "us-west-2";
```

The output will be something like this:

```
[
    {
        "IdOfTheSecurityGroup": "sg-049e93c5df3e056dc",
        "IdOfTheVPC": "vpc-00bc84d9f66dd56f0"
    },{
        "IdOfTheSecurityGroup": "sg-09cfc574cf19a85cc",
        "IdOfTheVPC": "vpc-06b64f761be00fcb3"
```

```
        }
    ]
```

I would be using an AWS CloudShell to SSH into the EC2 instance using the public IPv4 address. To find the public IP address of the EC2 instance, let's use the dig command.

This package is not installed by default on CloudShell, but it can be installed, of course. Just use this command to install in silent mode suppressing confirmations:

```
sudo yum install bind-utils
```

To find the public IP address for your CloudShell session, type the following:

```
dig TXT +short o-o.myaddr.1.google.com @ns1.google.com
```

My simple output is "34.212.40.234".

Let's modify the security group to allow incoming SSH requests originating from this IPv4 address. Here is the command:

```
aws ec2 authorize-security-group-ingress \
    --group-id sg-049e93c5df3e056dc \
    --protocol tcp \
    --port 22 \
    --cidr 34.212.40.234/32 \
    --region "us-west-2";
```

The output will be something like this:

```
{
    "Return": true,
    "SecurityGroupRules": [
        {
            "SecurityGroupRuleId": "sgr-08fee0cd337d2c300",
            "GroupId": "sg-049e93c5df3e056dc",
            "GroupOwnerId": "123456789012",
            "IsEgress": false,
            "IpProtocol": "tcp",
            "FromPort": 22,
            "ToPort": 22,
            "CidrIpv4": "34.212.40.234/32"
```

```
            }
      ]
}
```

Just repeat the command changing the -group-id parameters accordingly to change the security group for VPC [01]. We are almost there; we are done configuring the network and now it is finally time to launch our EC2 instance. We decided to select one, m5d.12xlarge, to show how to use the **non-volatile memory express** (**NVMe**) instance store. Grab a coffee and relax, there are more CLI commands coming your way!

Outposts instance launch example

Let's begin by selecting an **Amazon Machine Image** (**AMI**). To make things simpler, I will use Amazon-provided AMIs and not from the Marketplace. I will choose an Ubuntu Server image because it is a popular Linux distro. NVMe support should be better on newer versions, therefore, I will choose among the variants of Ubuntu 20.04 running SQL Server, a database application that will love fast storage. Whenever you're ready, type the following:

```
aws ec2 describe-images \
    --filters
"Name=description,Values=Ubuntu*Server*20.04*SQL*2019*Express
*Amazon*" \
    --query "Images[].
[CreationDate,ImageId,Description,ImageOwnerAlias]" \
    --region "us-west-2";
```

The output will be something like this:

```
[
    [
        "2022-05-02T23:55:34.000Z",
        "ami-01965e543ed9274e1",
        "Ubuntu Server 20.04 with SQL Server 2019 Express
Edition AMI provided by Amazon.",
        "amazon"
    ],[
        "2021-05-02T23:24:53.000Z",
        "ami-03f9c561d1e164015",
        "Ubuntu Server 20.04 with SQL Server 2019 Express
Edition AMI provided by Amazon.",
        "amazon"
```

```
    ], [
        "2021-09-24T19:54:09.000Z",
        "ami-09fd3d01ba256804e",
        "Ubuntu Server 20.04 with SQL Server 2019 Express
Edition AMI provided by Amazon.",
        "amazon"
    ], [
        "2022-02-05T00:13:52.000Z",
        "ami-0a049ac60d3caadd2",
        "Ubuntu Server 20.04 with SQL Server 2019 Express
Edition AMI provided by Amazon.",
        "amazon"
    ]
]
```

The most recent AMI should probably be the most up-to-date. Using this criterion, the winner is AMI ID ami-01965e543ed9274e1. One prerequisite prior to launching our instance is to create one EC2 key pair to authenticate our SSH request. Okay, here is the command:

```
aws ec2 create-key-pair –key-name OP1.LAB01 –key-format pem –
query 'KeyMaterial' –output text > kp-OP1.LAB01.pem
```

The output of this command created the kp-OP1.LAB01.pem file on the current directory of our CloudShell session. There is a documentation page describing this process, which can be found at this URL: https://docs.aws.amazon.com/cli/latest/userguide/cli-services-ec2-keypairs.html. One step described at this URL is one Linux bash command to modify the security for the .pem file. Here it is:

```
chmod 400 kp-OP1.LAB01.pem
```

You need to understand how to select the NVMe storage and how mappings are performed inside EC2; please check this documentation: https://docs.aws.amazon.com/AWSEC2/latest/UserGuide/block-device-mapping-concepts.html. In short, ephemeral storage is mapped inside Linux to the /dev/sdb device, ephemeral[0-3].

Let's recap the parameters for our launch:

- AMI ID: ami-01965e543ed9274e1
- Instance type: m5d.12xlarge
- Block device mapping: /dev/sdb, ephemeral0
- Subnet ID: subnet-02a96f65e7a82ee7c (vpc02_cidr2_usw2-az1_subnet04)

- Key pair name: `OP1.LAB01`

- Security group ID: `sg-049e93c5df3e056dc` (default security group for VPC [02])

Here is the command for the magic to happen:

```
aws ec2 run-instances \
    --image-id ami-01965e543ed9274e1 \
    --count 1 \
    --instance-type m5d.12xlarge \
    --block-device-mappings "DeviceName=/dev/
sdb,VirtualName=ephemeral0" \
    --subnet-id subnet-02a96f65e7a82ee7c \
    --key-name "OP1.LAB01" \
    --security-group-ids sg-049e93c5df3e056dc
    --associate-public-ip-address \
    --region "us-west-2";
```

Take a break to digest the full output:

```
{
    "Groups": [],
    "Instances": [
        {
            "AmiLaunchIndex": 0,
            "ImageId": "ami-01965e543ed9274e1",
            "InstanceId": "i-0c9a547400506f362",
            "InstanceType": "m5d.12xlarge",
            "KeyName": "OP1.LAB01",
            "LaunchTime": "2022-05-16T01:20:40+00:00",
            "Monitoring": {
                "State": "disabled"
            },
            "Placement": {
                "AvailabilityZone": "us-west-2a",
                "GroupName": "",
                "Tenancy": "default"
            },
            "PrivateDnsName": "ip-10-20-55-232.us-west-2.
```

```
compute.internal",
            "PrivateIpAddress": "10.20.55.232",
            "ProductCodes": [],
            "PublicDnsName": "",
            "State": {
                "Code": 0,
                "Name": "pending"
            },
            "StateTransitionReason": "",
            "SubnetId": "subnet-02a96f65e7a82ee7c",
            "VpcId": "vpc-00bc84d9f66dd56f0",
            "Architecture": "x86_64",
            "BlockDeviceMappings": [],
            "ClientToken": "804416e8-05c9-41d8-a588-
b6c4db93c343",
            "EbsOptimized": false,
            "EnaSupport": true,
            "Hypervisor": "xen",
            "NetworkInterfaces": [
                {
                    "Attachment": {
                        "AttachTime": "2022-05-
16T01:20:40+00:00",
                        "AttachmentId": "eni-attach-
0bc627f2b9fe52da4",
                        "DeleteOnTermination": true,
                        "DeviceIndex": 0,
                        "Status": "attaching",
                        "NetworkCardIndex": 0
                    },
                    "Description": "",
                    "Groups": [
                        {
                            "GroupName": "default",
                            "GroupId": "sg-049e93c5df3e056dc"
                        }
                    ],
```

```
                "Ipv6Addresses": [],
                "MacAddress": "02:e7:64:e7:ed:63",
                "NetworkInterfaceId": "eni-
0c25dd8cbd5373eb3",
                "OwnerId": "123456789012",
                "PrivateIpAddress": "10.20.55.232",
                "PrivateIpAddresses": [
                    {
                        "Primary": true,
                        "PrivateIpAddress": "10.20.55.232"
                    }
                ],
                "SourceDestCheck": true,
                "Status": "in-use",
                "SubnetId": "subnet-02a96f65e7a82ee7c",
                "VpcId": "vpc-00bc84d9f66dd56f0",
                "InterfaceType": "interface"
            }
        ],
        "OutpostArn": "arn:aws:outposts:us-west-
2:123456789012:outpost/op-8b286039iedad23e0",
        "RootDeviceName": "/dev/sda1",
        "RootDeviceType": "ebs",
        "SecurityGroups": [
            {
                "GroupName": "default",
                "GroupId": "sg-049e93c5df3e056dc"
            }
        ],
        "SourceDestCheck": true,
        "StateReason": {
            "Code": "pending",
            "Message": "pending"
        },
        "VirtualizationType": "hvm",
        "CpuOptions": {
            "CoreCount": 24,
```

```
                    "ThreadsPerCore": 2
                },
                "CapacityReservationSpecification": {
                    "CapacityReservationPreference": "open"
                },
                "MetadataOptions": {
                    "State": "pending",
                    "HttpTokens": "optional",
                    "HttpPutResponseHopLimit": 1,
                    "HttpEndpoint": "enabled",
                    "HttpProtocolIpv6": "disabled",
                    "InstanceMetadataTags": "disabled"
                },
                "EnclaveOptions": {
                    "Enabled": false
                },
                "PrivateDnsNameOptions": {
                    "HostnameType": "ip-name",
                    "EnableResourceNameDnsARecord": false,
                    "EnableResourceNameDnsAAAARecord": false
                },
                "MaintenanceOptions": {
                    "AutoRecovery": "default"
                }
            }
        ],
        "OwnerId": "123456789012",
        "ReservationId": "r-04c54feac677d46f9"
    }
```

As before, let's check some key parameters:

- InstanceId: i-0c9a547400506f362

- PrivateIpAddress: 10.20.55.232

- OutpostArn: arn:aws:outposts:us-west-2:123456789012:outpost/op-8b286039iedad23e0

Happiness! We are live, running inside AWS Outposts. We just need to wait for our instance to achieve the **Running** state and pass status checks. How can we see this? Well, it is simpler than you could suppose; the command is here:

```
aws ec2 describe-instance-status \
    --instance-ids i-0c9a547400506f362 \
    --region "us-west-2";
```

Here is the output:

```
{
    "InstanceStatuses": [
        {
            "AvailabilityZone": "us-west-2a",
            "OutpostArn": "arn:aws:outposts:us-west-
2:123456789012:outpost/op-8b286039iedad23e0",
            "InstanceId": "i-0c9a547400506f362",
            "InstanceState": {
                "Code": 16,
                "Name": "running"
            },
            "InstanceStatus": {
                "Details": [
                    {
                        "Name": "reachability",
                        "Status": "passed"
                    }
                ],
                "Status": "ok"
            },
            "SystemStatus": {
                "Details": [
                    {
                        "Name": "reachability",
                        "Status": "passed"
                    }
                ],
                "Status": "ok"
```

```
            }
        }
    ]
}
```

Did you miss something when we received the **JSON** output confirming the instance launch? Yes, me too. That output did not show the public IPv4 address that should have been automatically assigned. This behavior is expected, but fear nothing: AWS CLI to the rescue once again! Just type this command:

```
aws ec2 describe-instances \
    --instance-ids i-0c9a547400506f362 \
    --query "Reservations[].Instances[].
{ConnectionIpAddress:PublicIpAddress}" \
    --region "us-west-2";
```

The output is as follows:

```
[
    {
        "ConnectionIpAddress": "34.217.64.85"
    }
]
```

Can you feel it? This is the moment we have been working so hard to accomplish, the time to finally connect to our instance; more details on connecting to EC2 instances can be found at this URL: https://docs.aws.amazon.com/AWSEC2/latest/UserGuide/AccessingInstancesLinux.html.

But if you can't wait any longer, hold your breath and type the following:

```
ssh -i kp-OP1.LAB01.pem ubuntu@34.217.64.85
```

Hopefully, you are now smiling and as grateful as I am. If you see this output, congratulations! You endured this journey and it was a success!

```
[cloudshell-user@ip-10-0-161-251 ~]$ ssh -i kp-OP1.LAB01.pem
ubuntu@34.217.64.85
The authenticity of host '34.217.64.85 (34.217.64.85)' can't be
established.
ECDSA key fingerprint is
SHA256:HyHaCX1X3AWjT+AhUP1Aujn2xII63kRmC+fFABhlEKY.
ECDSA key fingerprint is MD5:11:56:13:7d:26:3a:f1:a4:77:9a:7b:8
```

```
2:57:5c:60:43.
Are you sure you want to continue connecting (yes/no)? yes
Warning: Permanently added '34.217.64.85' (ECDSA) to the list
of known hosts.
Welcome to Ubuntu 20.04.4 LTS (GNU/Linux 5.13.0-1022-aws
x86_64)
```

Check your instance metadata with this simple command:

```
ubuntu@ip-10-20-55-232:~$ ec2metadata
```

This is my output:

```
ami-id: ami-01965e543ed9274e1
ami-launch-index: 0
ami-manifest-path: (unknown)
ancestor-ami-ids: unavailable
availability-zone: us-west-2a
block-device-mapping: ami
ephemeral0
ephemeral1
root
instance-action: none
instance-id: i-0c9a547400506f362
instance-type: m5d.12xlarge
local-hostname: ip-10-20-55-232.us-west-2.compute.internal
local-ipv4: 10.20.55.232
kernel-id: unavailable
mac: unavailable
profile: default-hvm
product-codes: unavailable
public-hostname: unavailable
public-ipv4: 34.217.64.85
public-keys: ['ssh-rsa AAA3NzaC1yc2EAABAQCoXEn1EXvAQABEW/
jf9k2G63WSQe9SmC4BcJ30ABeIkqpcYAKVAQ/wrS6/
nOPmEx1W01kN7hSLhCIVvmy17Pq5gVlNYAn1Ad7exHrrSnlNFmbFcZVfyr2WTxX
EXCI4vqM5FO1HSAQAJNsMzL2ywIS6RYCr02zDZ1q+Um1adrXGv7Ch98HCT60XS
POpMZv18tO+FJw5Gi3SIHPXP41Hg3OfDQgEULWpy8gH8U11t6vOdjRPQk7KQeC8
xmADApknFXHFeIPn44AYc2LuEaM9smCAUd1SH2MQc3jSSeRezcVxQYoLRpOVdU
```

```
gpVYr+LxlxJXKSyKXgSpkT0Yp2ZtnbSCaZ  sWh1QR1FlKre  OP1.LAB01']
ramdisk-id: unavailable
reservation-id: r-04c54feac677d46f9
security-groups: default
user-data: unavailable
```

What about our NVMe disks? Here they are; type the following:

```
sudo apt update
sudo apt install nvme-cli
sudo nvme list
```

Here is my output:

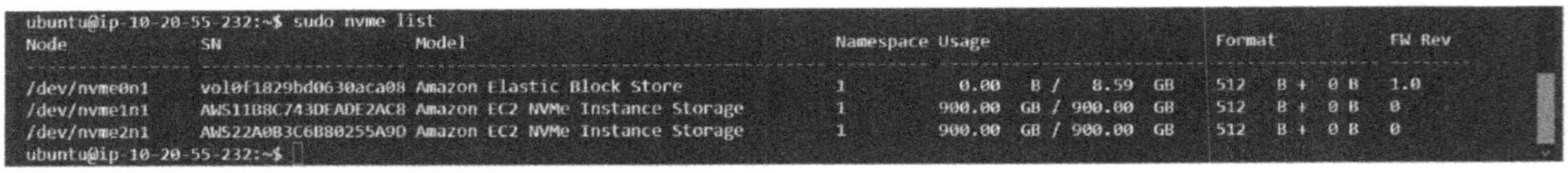

Figure 4.7 – NVMe volumes attached to the m5d.12xlarge instance

Before we wrap up, we'll have one more play with our instance. How many vCPUs does it have? Type the following:

```
ubuntu@ip-10-20-55-232:~$ nproc --all
```

My result is 48.

Here's another tidbit to find out more about the vCPUs:

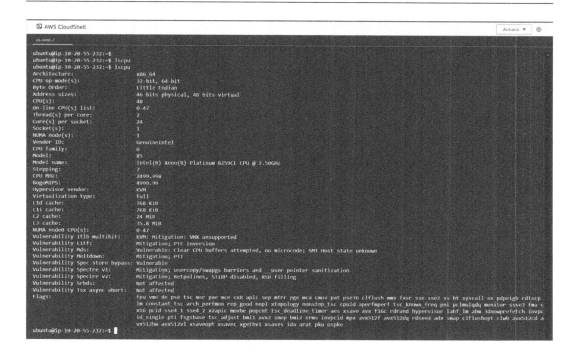

Figure 4.8 – CPU features for the m5d.12xlarge instance

One more check, this time the memory features:

```
cat /proc/meminfo
```

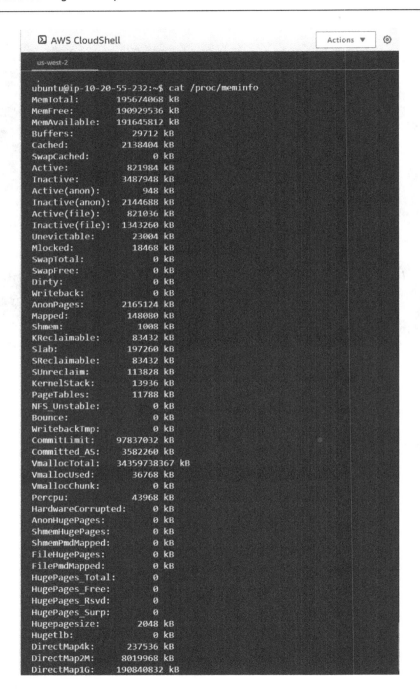

Figure 4.9 – Memory features for the m5d.12xlarge instance

To wrap up, if you are exercising *Learn and be curious*, you are probably thinking, *What about the SQL Server?* Let's wrap up by unveiling the status of the SQL Server service. Type the following:

```
sudo systemctl mssql-server.service status
```

```
ubuntu@ip-10-20-55-232:~$ sudo systemctl mssql-server.service status
Unknown operation mssql-server.service.
ubuntu@ip-10-20-55-232:~$ sudo systemctl status mssql-server.service
● mssql-server.service - Microsoft SQL Server Database Engine
     Loaded: loaded (/lib/systemd/system/mssql-server.service; enabled; vendor preset: enabled)
     Active: active (running) since Mon 2022-05-16 01:21:40 UTC; 1h 14min ago
       Docs: https://docs.microsoft.com/en-us/sql/linux
   Main PID: 965 (sqlservr)
      Tasks: 245
     Memory: 3.0G
     CGroup: /system.slice/mssql-server.service
             ├─ 965 /opt/mssql/bin/sqlservr
             └─3237 /opt/mssql/bin/sqlservr

May 16 01:25:06 ip-10-20-55-232 sqlservr[3237]: [124B blob data]
May 16 01:25:06 ip-10-20-55-232 sqlservr[3237]: [143B blob data]
May 16 01:25:06 ip-10-20-55-232 sqlservr[3237]: [131B blob data]
May 16 01:25:06 ip-10-20-55-232 sqlservr[3237]: [143B blob data]
May 16 01:25:06 ip-10-20-55-232 sqlservr[3237]: [157B blob data]
May 16 01:25:06 ip-10-20-55-232 sqlservr[3237]: [193B blob data]
May 16 01:25:06 ip-10-20-55-232 sqlservr[3237]: [155B blob data]
May 16 01:25:07 ip-10-20-55-232 sqlservr[3237]: [204B blob data]
May 16 01:30:06 ip-10-20-55-232 sqlservr[3237]: [156B blob data]
May 16 01:30:06 ip-10-20-55-232 sqlservr[3237]: [195B blob data]
ubuntu@ip-10-20-55-232:~$
```

Figure 4.10 – SQL Server status features for the m5d.12xlarge instance

We made it! In this section, we identified our Outposts capabilities and its parent AZ. Next, we implemented a comprehensive network design with multiple VPCs and multiple subnets associated with multiple AZs. Finally, we created an EC2 key pair, identified AMIs to use, and connected our AWS CloudShell to an EC2 instance launched inside AWS Outposts via SSH.

In the next section, we will discover how to work with this distinct construct specific to AWS Outposts, the local gateway. We will discover its objects, parameters, and configurations and how we can distinguish traffic destined to on-premises networks from packets sent to the Region via a service link.

Working with local gateways

Our EC2 instance is now running on AWS Outposts and is capable of communicating with the other subnets in VPC and also with the internet because it has a public IPv4 associated with it. So, what role does the local gateway play?

The **local gateway** (**LGW**) is the networking construct that allows your Outpost to communicate with the customer network, and it only exists in the rack. Outposts servers use the **Local Network Interface** (**LNI**) to communicate with the local network.

Only one LGW is created per Outpost and it can be attached to multiple VPCs within the Outpost; it operates in a **Network Address Translation (NAT)** fashion. This feature can add additional latency to the network traffic in extreme conditions. It's a best practice to interact with AWS teams to assess the limitations of this structure under stress conditions and make sure it will meet your architecture requirements.

During the installation process, another type of route table is deployed and this object is specifically associated with your AWS Outpost LGW to enable packets to be forwarded to the customer network. This structure is denominated **local gateway route table**. What if we put the AWS CLI to work once again to describe these structures in our rack? Hold on tight, let's go!

Local gateway structure

Let's begin describing our LGW. Just type the following:

```
aws ec2 describe-local-gateways \
    --filters "Name=outpost-arn,Values=arn:aws:outposts:us-
west-2:123456789012:outpost/op-8b286039iedad23e0" \
    --region "us-west-2";
```

The output should be something like this:

```
{
    "LocalGateways": [
        {
            "LocalGatewayId": "lgw-0cdc67d1ae6c75ff8",
            "OutpostArn": "arn:aws:outposts:us-west-
2:123456789012:outpost/op-8b286039iedad23e0",
            "OwnerId": "123456789012",
            "State": "available",
            "Tags": []
        }
    ]
}
```

Let's make note of the retrieved local gateway ID, `lgw-0cdc67d1ae6c75ff8`, and describe the LGW route table. Here is the command:

```
aws ec2 describe-local-gateway-route-tables \
    --filters "Name=local-gateway-id,Values=lgw-
```

```
Ocdc67d1ae6c75ff8" \
    --region "us-west-2";
```

The output should be similar to this:

```
{
    "LocalGatewayRouteTables": [
        {
            "LocalGatewayRouteTableId": "lgw-rtb-
0aeee96ea969f0f25",
            "LocalGatewayRouteTableArn": "arn:aws:ec2:us-
west-2:123456789012:local-gateway-route-table/lgw-rtb-
0aeee96ea969f0f25",
            "LocalGatewayId": "lgw-0cdc67d1ae6c75ff8",
            "OutpostArn": "arn:aws:outposts:us-west-
2:123456789012:outpost/op-8b286039iedad23e0",
            "OwnerId": "123456789012",
            "State": "available",
            "Tags": [
                {
                    "Key": "Name",
                    "Value": " IT Datacenter 02 in Oregon
Outpost 02 AZ01"
                }
            ]
        }
    ]
}
```

Local gateway route table

Naturally, we expect route tables to have routes associated. How do we find them for this special type of route table designed specifically for the local gateway? There is a specific API call to perform this operation; we need the LGW route table ID, `lgw-rtb-0aeee96ea969f0f25`, retrieved in the last call. Here it is:

```
aws ec2 search-local-gateway-routes \
    --local-gateway-route-table-id "lgw-rtb-0aeee96ea969f0f25"
\
    --region "us-west-2";
```

Here is the output:

```
{
    "Routes": [
        {
            "DestinationCidrBlock": "0.0.0.0/0",
            "LocalGatewayVirtualInterfaceGroupId": "lgw-vif-
grp-033d2b33464749f3a",
            "Type": "static",
            "State": "active",
            "LocalGatewayRouteTableId": "lgw-rtb-
0aeee96ea969f0f25",
            "LocalGatewayRouteTableArn": "arn:aws:ec2:us-
west-2:123456789012:local-gateway-route-table/lgw-rtb-
0aeee96ea969f0f25",
            "OwnerId": "123456789012"
        }
    ]
}
```

The preceding output reveals a new structure identified as LocalGatewayVirtualInterfaceGroupId, with the value of lgw-vif-grp-033d2b33464749f3a. This object represents the virtual interfaces that are associated with parameters set on **Outpost Networking Devices** (**ONDs**) allowing traffic to the customer network, and note the default route (0.0.0.0/0) associated with this object, indicating that all traffic that doesn't match specific routes in the routing table must be forwarded to the local network.

This is generally a good practice; traffic being forwarded to the region by default can saturate the service link or accrue additional costs as data transfer charges.

During the provisioning phase of the rack, the OND is configured with a set of parameters needed to initiate a **Border Gateway Protocol** (**BGP**) session with the **Customer Network Device** (**CND**) and begin to exchange routes. These are as follows:

- VLAN number
- Subnet CIDR block
- Outposts IP
- Customer IP
- Customer ASN

As usual, we have API calls to describe these interfaces and parameters. Let's begin with the interfaces:

```
aws ec2 describe-local-gateway-virtual-interface-groups \
    --local-gateway-virtual-interface-group-ids "lgw-vif-grp-
033d2b33464749f3a" \
    --region "us-west-2";
```

Here is what the output should look like:

```
{
    "LocalGatewayVirtualInterfaceGroups": [
        {
            "LocalGatewayVirtualInterfaceGroupId": "lgw-vif-
grp-033d2b33464749f3a",
            "LocalGatewayVirtualInterfaceIds": [
                "lgw-vif-08a0463925b3b3018",
                "lgw-vif-0e3d61695e0853631"
            ],
            "LocalGatewayId": "lgw-0cdc67d1ae6c75ff8",
            "OwnerId": "123456789012",
            "Tags": []
        }
    ]
}
```

We can identify two LGW virtual interfaces, lgw-vif-08a0463925b3b3018 and lgw-vif-0e3d61695e0853631. Here is the API call to describe the interface parameters:

```
aws ec2 describe-local-gateway-virtual-interfaces \
    --filters "Name=local-gateway-id,Values=lgw-
0cdc67d1ae6c75ff8" \
    --region "us-west-2";
```

Here is the output:

```
{
    "LocalGatewayVirtualInterfaces": [
        {
            "LocalGatewayVirtualInterfaceId": "lgw-vif-
08a0463925b3b3018",
```

```
            "LocalGatewayId": "lgw-0cdc67d1ae6c75ff8",
            "Vlan": 2406,
            "LocalAddress": "169.254.6.0/31",
            "PeerAddress": "169.254.6.1/31",
            "LocalBgpAsn": 65006,
            "PeerBgpAsn": 65000,
            "OwnerId": "123456789012",
            "Tags": []
        },
        {
            "LocalGatewayVirtualInterfaceId": "lgw-vif-
0e3d61695e0853631",
            "LocalGatewayId": "lgw-0cdc67d1ae6c75ff8",
            "Vlan": 2406,
            "LocalAddress": "169.254.6.2/31",
            "PeerAddress": "169.254.6.3/31",
            "LocalBgpAsn": 65006,
            "PeerBgpAsn": 65000,
            "OwnerId": "123456789012",
            "Tags": []
        }
    ]
}
```

In the output, we can identify the parameters to be set on the Outposts side (local) and the expected parameters to be configured at the other end of the connection on the CND (peer). One VLAN will carry the service link traffic and the other will be responsible for forwarding traffic designated to the customer's local network, thus logically segregating the data flow.

To enable instances launched inside AWS Outposts to communicate with the customer's local network, we have to associate the VPCs that contain subnets created inside Outposts with the LGW route table. If you don't perform this association, EC2 instances running inside Outposts can talk to the Region but not talk to services and processes running in the local network.

An important point: you can associate multiple VPCs with your local gateway route table, but CIDR blocks among VPCs must not overlap. You will receive an error when you try to associate the VPC with the local gateway route table. Let's describe the VPC associations:

```
aws ec2 describe-local-gateway-route-table-vpc-associations \
    --local-gateway-route-table-vpc-association-ids \
```

```
    --filters "Name=local-gateway-id,Values=lgw-
0cdc67d1ae6c75ff8" \
    --region "us-west-2";
```

Here is the output:

```
{
    "LocalGatewayRouteTableVpcAssociations": [
        {
            "LocalGatewayRouteTableVpcAssociationId": "lgw-vpc-
assoc-00524c139044d334c",
            "LocalGatewayRouteTableId": "lgw-rtb-
0aeee96ea969f0f25",
            "LocalGatewayRouteTableArn": "arn:aws:ec2:us-
west-2:123456789012:local-gateway-route-table/lgw-rtb-
0aeee96ea969f0f25",
            "LocalGatewayId": "lgw-0cdc67d1ae6c75ff8",
            "VpcId": "vpc-06b64f761be00fcb3",
            "OwnerId": "123456789012",
            "State": "associated",
            "Tags": []
        },{
            "LocalGatewayRouteTableVpcAssociationId": "lgw-vpc-
assoc-00cde271be2808da9",
            "LocalGatewayRouteTableId": "lgw-rtb-
0aeee96ea969f0f25",
            "LocalGatewayRouteTableArn": "arn:aws:ec2:us-
west-2:123456789012:local-gateway-route-table/lgw-rtb-
0aeee96ea969f0f25",
            "LocalGatewayId": "lgw-0cdc67d1ae6c75ff8",
            "VpcId": "vpc-00bc84d9f66dd56f0",
            "OwnerId": "123456789012",
            "State": "associated",
            "Tags": []
        }
    ]
}
```

The final step to communicate with services and processes running on the on-premises network would be to associate one elastic IP allocated from the **customer-owned IP address pool (CoIP pool)** that is configured during Outposts' initial installation. If you need additional pools, you will need to open a support case with AWS.

The equipment for this lab was not associated with a CoIP but I will explain what needs to be done when you get yours, which will certainly have the CoIP range that you specified. On the AWS console EC2 page, you will find the **Elastic IPs** menu, and you just need to select **Allocate Elastic IP address**:

Figure 4.11 – AWS Console EC2 service – Elastic IPs

On the **Allocate Elastic IP address** page, you will have the option to select a customer-owned pool of IPv4 addresses as the source. Make your selection, click **Allocate**, and you're done. Your instance should now be able to communicate both ways, with the Region via the service link and with the on-premises network via the LGW.

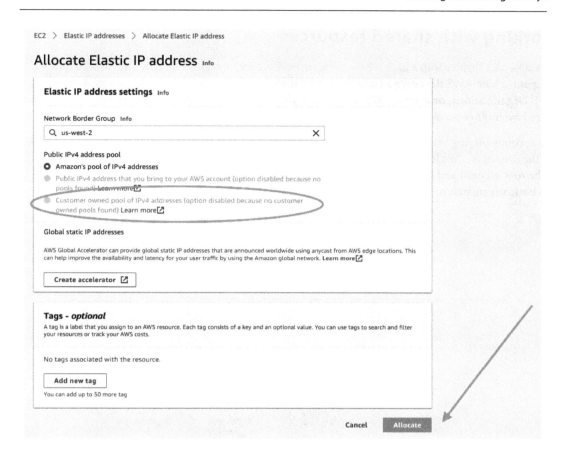

Figure 4.12 – AWS Console EC2 service – Allocate Elastic IPs

There is an AWS blog describing this process with more detail for a specific use case at this URL: `https://aws.amazon.com/blogs/storage/connecting-aws-outposts-to-on-premises-data-sources/`.

That's it, with an elastic IP from the customer-owned IP pool associated with our instance, it is now able to send and receive packets to the on-premises customer network. If you want to assign one specific IP address, you will need to use the AWS CLI. If you use the AWS Management Console, one IP will be randomly assigned from the pool of unused IP addresses.

In the next section, we will learn how to share Outposts resources with other accounts and grant permissions for principals to use these resources.

Working with shared resources

Let's close this chapter with a look at how we share resources across different accounts. AWS Outposts is compatible with AWS **Resource Access Manager** (**RAM**), a service that securely allows AWS accounts, AWS Organizations, or **organizational units** (**OUs**) within an organization to consume resources shared by another account. Better yet, AWS RAM is available to you at no cost.

The account sharing the resources is denoted as the *owner*, while the accounts using these resources are the *consumers*. With this model, you can design one AWS Organization and order your Outposts in the root account and share its resources with OUs or member accounts directly. Here is an example of a basic organization:

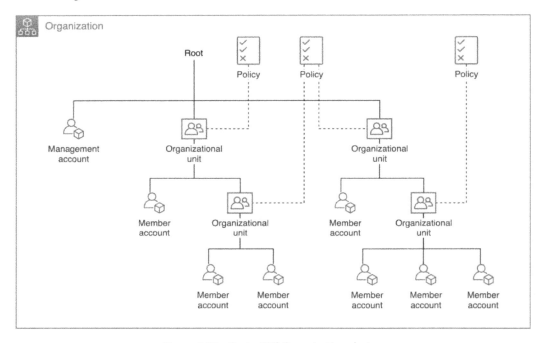

Figure 4.13 – Basic AWS Organization design

There are services that can run on Outposts but cannot be shared. A comprehensive list of resources that can be shared is available at this URL: https://docs.aws.amazon.com/outposts/latest/userguide/sharing-outposts.html#sharing-resources.

Implementing a structured AWS Organization tree to manage multiple accounts and enforce logical isolation, security boundaries, and role-based access control is a strongly advised best practice. It is referenced in the Well-Architected Framework's security, reliability, operational excellence, and cost optimization pillars.

With an established AWS Organization, you just need to enable sharing in the organization. You can perform this action using the AWS Console; just head on to the AWS RAM service landing page:

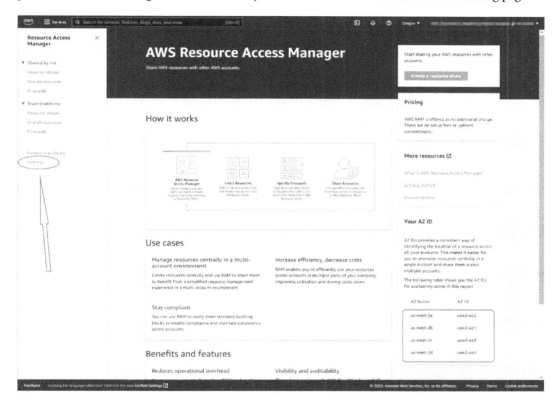

Figure 4.14 – AWS Resource Access Manager landing page

We know that AWS Outposts is anchored to just one AZ in the parent region and that AWS abstracts the AZ ID using the AZ name to spread the resource allocation more evenly. For this reason, we need to identify the AZ ID, which is the real AZ identification used internally at AWS, to its corresponding AZ name, which is the logical AZ identification randomly assigned to each AWS customer account and informed in the AWS Console.

This mapping is shown on the landing page in the **Your AZ ID** section. To enable sharing, click **Settings** on the left pane, as shown in *Figure 4.14*.

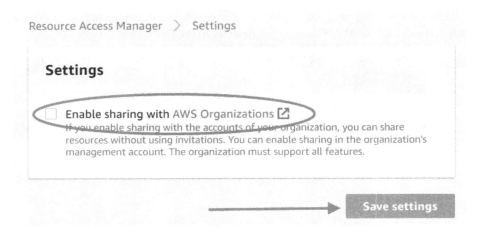

Figure 4.15 – AWS Resource Access Manager – Settings

You just need to check **Enable sharing within your AWS Organization** and click **Save settings**.

The **Resource shares** menu entry on the left pane shows the active resource shares for this account. These shares contain one or more resource IDs and the permissions given to principals who were granted access to these resources:

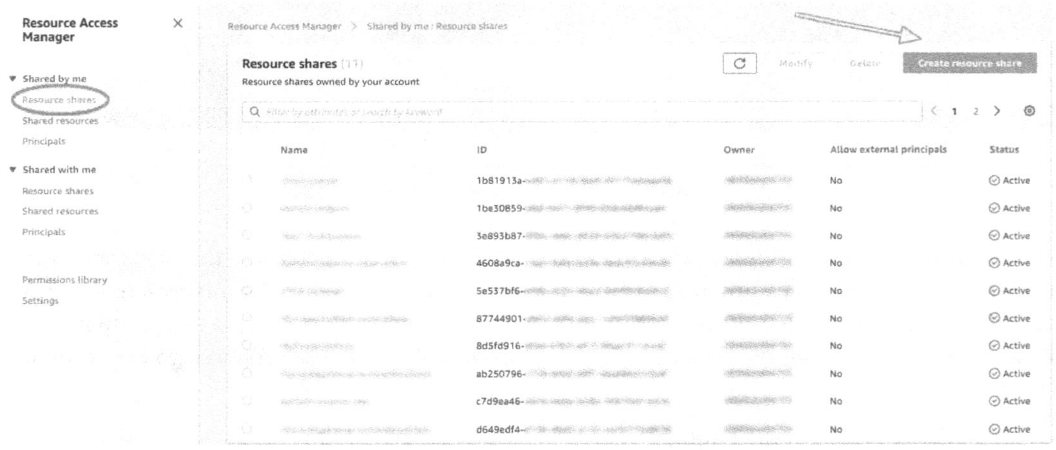

Figure 4.16 – AWS Resource Access Manager – Resource shares

Now, let's demonstrate how to create a resource share. Click on **Create resource share**, as indicated in *Figure 4.16*. This will take us to the first step of the **Specify resource share details** process. On this page, we specify **Resource share name** and select the resources to be shared. Here, you can define **Tags**, which is always a best practice. Click **Next**:

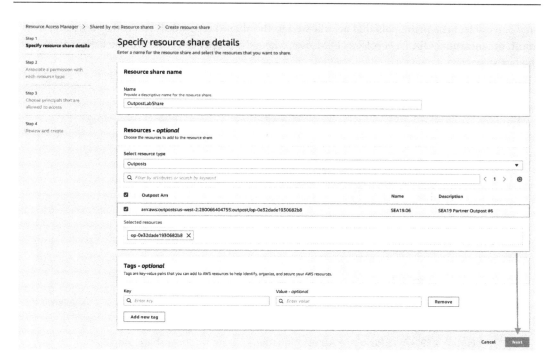

Figure 4.17 – AWS Resource Access Manager – Create resource share, step 1

Next, you will define permissions granted to the principals who will be able to use the shared resource. Click **Next**:

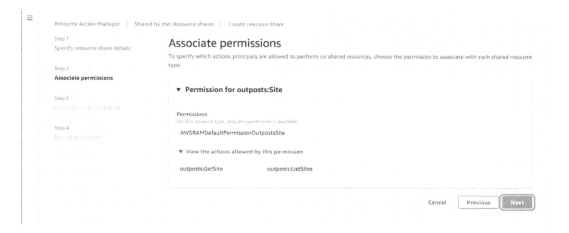

Figure 4.18 – AWS Resource Access Manager – Create resource share, step 2

In *step 3*, we select the principals that are allowed to the shared resources. You can specify an AWS account, organization, OU, IAM role, or IAM user. Make the proper selection and click **Next** to move to the last step:

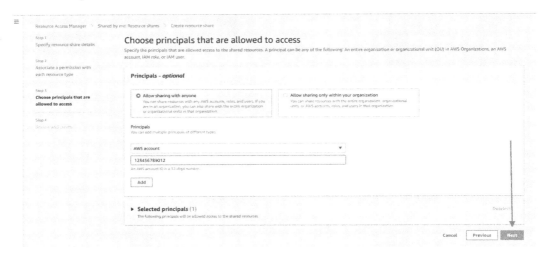

Figure 4.19 – AWS Resource Access Manager – Create resource share, step 3

In *step 4*, you will review your selections and, finally, create your share by clicking **Create resource share**:

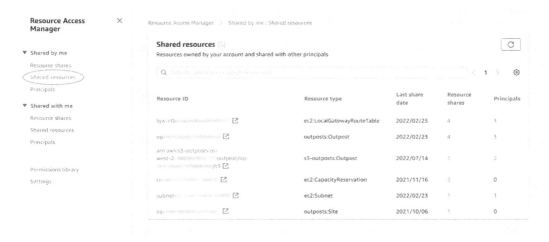

Figure 4.20 – AWS Resource Access Manager – Create resource share, step 4

On the service landing page on the left pane, the **Shared resources** menu entry shows the resource IDs owned by this account that were shared with other principals:

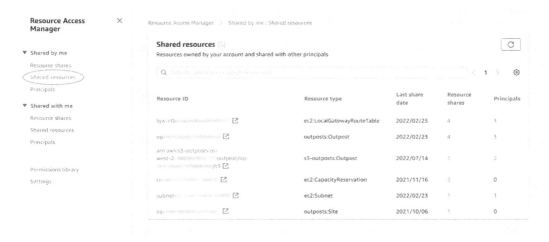

Figure 4.21 – AWS Resource Access Manager – Shared resources

Finally, the **Principals** menu entry shows the principals that your account has shared resources with:

Figure 4.22 – AWS Resource Access Manager – Principals

We have now finished the section about sharing resources. This is an important capability of Outposts, the ability to be used simultaneously by multiple accounts that can belong to different customers. It opens up the opportunity for a myriad of designs and use cases where Outposts can serve multiple tenants.

Remember, Outposts is not multi-tenant by default. It is your task to implement security boundaries and guardrails to prevent one tenant from having unauthorized access to another tenant's resources.

Summary

Congratulations on making it through the journey so far! This chapter was packed with information, but also an amazing experience for the hungry mind. We began with the challenging task of identifying our Outposts configurations and capabilities, implementing a complex network scenario from scratch, and launching an EC2 instance inside Outposts.

Next, we understood the local gateway, which is unique to AWS Outposts, and allowed it to exchange data with customer network segments. Finally, we learned how to share Outposts resources with other accounts.

You accomplished a lot, but there is more to come. In the next chapter, we will learn about security, and as they say at AWS: *Security is Job Zero*. We will talk about identity and access management, managed policies, physical security, and compliance. Security is paramount; we must strive to make security our top priority.

5

Security Aspects in Outposts

We finally had our hands-on session accessing the AWS Management Console to see the Outposts rack in action and we have established that AWS delivers on its promise to create an offering that operates as an extension of AWS in your data center.

If you have already invested in becoming an AWS professional, mastering the concepts of AWS Outposts will be smooth. However, if you are new to AWS, you will realize that AWS gives you the power to create apps operating at a massive scale for global deployment.

When operating within the confines of an AWS Region, you can rely on AWS to provide security. In this chapter, though, you will learn about the various aspects of AWS security when AWS Outposts is running in your data center.

The following topics will be covered in this chapter:

- AWS capabilities and data protection in Outposts
- IAM in Outposts
- The physical security of Outposts
- Resiliency and compliance

Data protection

We had a lot of fun putting AWS Outposts to work, checking configurations, creating a network design from scratch, and launching instances. While this is great, we must remember that security is said at AWS to be *job zero*, so important that we have dedicated an entire chapter to discussing Outposts' capabilities and security features.

As you would have learned to expect by now, AWS uses the same services and principles used in the Region to secure data stored in Outposts. Data must be secured in transit and at rest and data deleted must be completely non-recoverable. AWS is designed to meet the most stringent data compliance requirements with comprehensive capabilities complemented by AWS services available in the Region.

When it comes to security, one thing that customers must always remember is that no mechanism or system in the world will be able to protect you from a poorly designed and maintained **authentication, authorization, and accounting (AAA)** infrastructure. Any IT service should be designed around security and not the opposite. Security is a foundational core function; therefore, it should be built with the highest standards.

Let's define *encryption* by using the definition provided by the **National Institute of Standards and Technology (NIST)**, found at this URL: `https://csrc.nist.gov/glossary/term/encryption`. According to NIST, encryption can be described as:

> *Cryptographic transformation of data (called "plaintext") into a form (called "ciphertext") that conceals the data's original meaning to prevent it from being known or used. If the transformation is reversible, the corresponding reversal process is called "decryption," which is a transformation that restores encrypted data to its original state.*

The core AWS service for encryption is AWS **Key Management Service (KMS)**. The KMS service offers an infrastructure for you to control the cryptographic keys used to protect your data. KMS integrates with most other AWS services.

You can create keys and control the life cycle and permissions of your keys. This service offers centralized control and inventory, alongside the ability to track usage in combination with AWS CloudTrail. You can find more information about AWS KMS at this URL: `https://aws.amazon.com/kms/features/`.

Any process for persistent data begins with sending said data over some type of medium to be stored in the data repository backend. When you are sending data over the wire, traversing multiple networks controlled by multiple entities, you need to ensure this operation is not performed in the clear, which would allow a malicious third party to capture your data and access its content.

To protect your data, AWS suggests that you follow the AWS Well-Architected Framework *apply security at all layers* design principle, defined in the Security pillar. When applying this principle to your data in transit, you should consider employing a multi-level strategy; because AWS Outposts uses the same services and components present in an AWS Region, you benefit from the secure nature of these fundamental building blocks.

This multi-layered approach can be identified at multiple layers defined by the OSI model. AWS transparently encrypts traffic flowing between its data centers at the physical layer. At the network layer, **Virtual Private Clouds (VPCs)** offer the ability to encrypt intra-VPC and inter-VPC traffic between supported EC2 instances within a Region using AEAD algorithms with 256-bit encryption. You can find the list of supported instance types at this URL: `https://docs.aws.amazon.com/AWSEC2/latest/UserGuide/data-protection.html`.

At the application layer, AWS suggests you use the **Transport Layer Security (TLS)** protocol among peers, creating a secure HTTPS connection to perform your API requests. Additionally, you can explore services such as **AWS Certificate Manager (ACM)** to implement your PKI infrastructure and manage public and private SSL/TLS certificates. More information can be found at this URL: http://aws. amazon.com/certificate-manager.

AWS Outposts follows the approach to security through its service link. Outposts communicates with its parent Region via set VPN tunnels established over your communication infrastructure. You can increase your security posture by selecting the private connectivity option to avoid data traversing public networks.

Once data travels over channels and networks reaching the target repository endpoint, it will be persisted in non-volatile storage. Regardless of how long this data will remain stored, you must always consider encrypting this data at rest. When you are operating in the Region, encryption of your EBS volumes is optional; you can easily request a volume to be created with encryption enabled by selecting a master key that would be used for encryption.

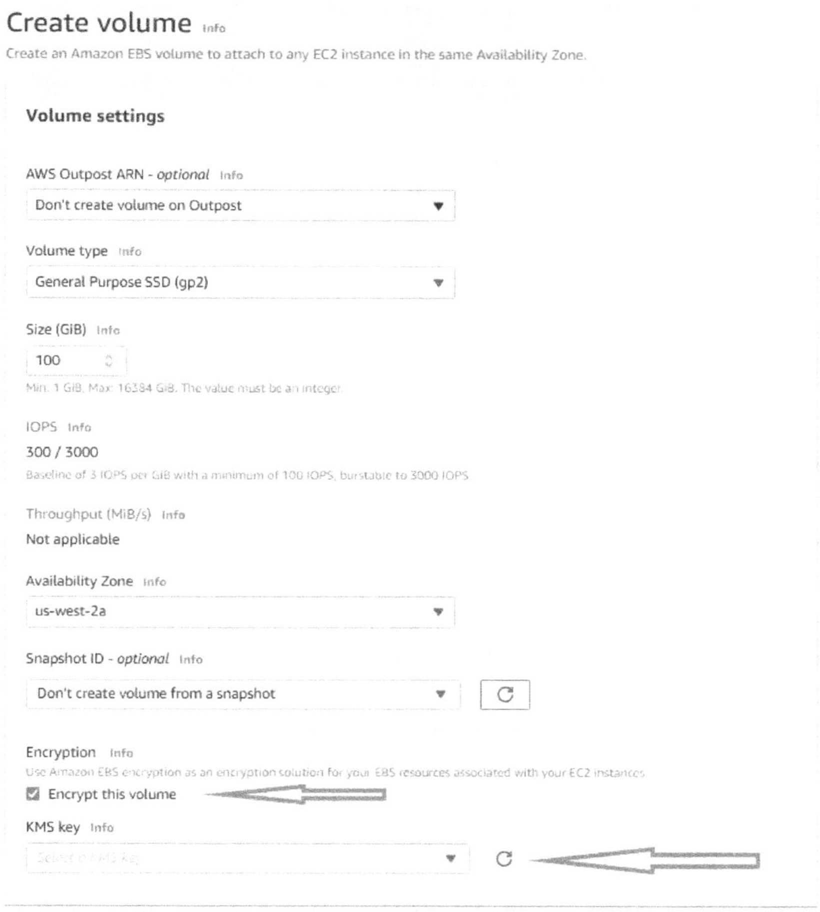

Figure 5.1 – AWS Management Console encryption options for EBS volumes

EC2 instances launched in AWS Outposts have their EBS volumes and snapshots encrypted by default. Even if you do not specify your own KMS key, the volume will be encrypted using an AWS managed key for EBS. You can find the key ID by heading to the AWS KMS service page and selecting **AWS managed keys** on the left pane.

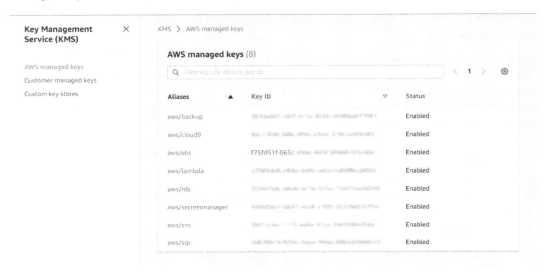

Figure 5.2 – AWS managed keys on the KMS service landing page

EBS volumes are secured with end-to-end encryption for any given data operation. Data transmission is secured during transit between the instance and the EBS volume. Snapshots created from this volume are also encrypted, as well as volumes created from those snapshots. The I/O latency penalty comparing encrypted and unencrypted volumes is negligible.

Eventually, when an instance is terminated, all the underlying data is securely disposed of. All storage blocks that make up a volume are reset and memory blocks allocated to the instance are all set to zero to scrub sensitive information.

If your hardware suffers an irrecoverable issue, AWS will send an instance retirement notice and schedule an AWS technician's onsite visit. The AWS technician will then replace the defective unit in its entirety, and no attempts will be made to replace components or diagnose the issue. Droplets will have their encryption module (**Nitro Security Key**, or **NSK**) removed and destroyed, effectively shredding any data remaining in the host. No data should leave your site!

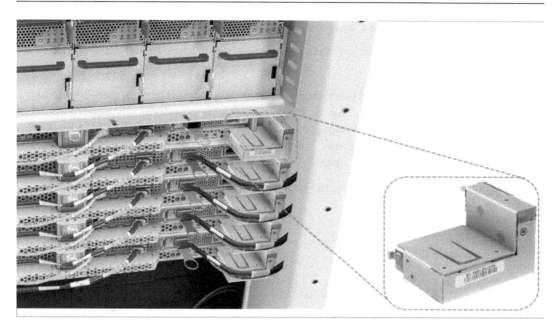

Figure 5.3 – Outposts NSK encryption module

Your data is the most valuable asset and it must be thoroughly protected by all possible means. You should never underestimate the potential of damage in the event of data leakage, especially if you handle **personally identifiable information (PII)**.

AWS provides extensive tooling around security, and Outposts by default enforces security best practices. If you believe *data is the new oil*, don't let your wealth be unprotected or exposed. In the next section, we will talk about how you can use the AWS IAM service to improve your security posture.

IAM in Outposts

Outposts uses the same services, components, and features available in the AWS cloud. It is a fundamental principle for the engineering of the product to develop Outposts as a native component of AWS infrastructure, integrating as seamlessly as if it were located inside an AWS Region.

The implementation of IAM mechanisms on Outposts leverages AWS IAM service as the building block to provide authentication and authorization. The same principles and concepts apply:

- **IAM resources**: Users, groups, roles, policies, and identity providers. These are the types of objects stored in IAM.

- **IAM entities**: Users and roles. You can use these elements to perform authentication. Users include federated users authenticated by other directories and roles assumed by other AWS accounts. The all-powerful entity is the root user, created when an AWS account is provisioned. These credentials should be secured and audited.

- **IAM identities**: Users, groups, and roles. You can attach policies to IAM identities.

- **IAM principals**: A person or application using IAM entities to sign and make requests to AWS resources.

By default, no IAM entities are granted permission to operate AWS Outposts or manage their resources. You have to explicitly attach a policy to the identity granting rights to perform actions on Outposts. I will provide one customer-managed example and you can find more details about IAM for AWS Outposts at this URL: https://docs.aws.amazon.com/outposts/latest/userguide/identity-access-management.html.

An example policy to grant *full read access* to Outposts is as follows:

```
{
    "Version": "2012-10-17",
    "Statement": [
        {
            "Effect": "Allow",
            "Action": [
                "outposts:ListTagsForResource",
                "outposts:ListSites",
                "outposts:ListOutposts",
                "cloudwatch:GetMetricData",
                "cloudwatch:ListMetricStreams",
                "outposts:GetOutpost",
                "outposts:GetOutpostInstanceTypes",
                "cloudwatch:GetMetricStatistics",
                "cloudwatch:ListMetrics"
            ],
            "Resource": "*"
        }
    ]
}
```

This policy is useful for auditing AWS Outposts. It allows all configurations and resources to be inspected but it does not allow changes. A professional will have access to all service parameters thus being able to describe the architecture at a great level of detail to perform a comprehensive analysis.

AWS services also need to interact and interoperate, performing calls from one to another. In this case, a special type of role comes into play.

Using service-linked roles

A person or application can use IAM identities to perform actions on AWS services and resources. AWS services are also applications, thereby being able to use a special type of IAM role that is directly linked to them.

These are called **service-linked** roles, whose purpose is to grant permissions to a given service to call other AWS services on your behalf. Outposts is listed as one of the services that can impersonate service-linked roles. You can find a comprehensive table of services at this URL: `https://docs.aws.amazon.com/IAM/latest/UserGuide/reference_aws-services-that-work-with-iam.html`.

One good example of a service that can use service-linked roles to perform an action on Outposts is Amazon RDS. This service is granted a policy called **AmazonRDSServiceRolePolicy** that grants it permission to execute actions on behalf of database instances on Outposts resources. You can find more details at this URL: `https://docs.aws.amazon.com/AmazonRDS/latest/UserGuide/UsingWithRDS.IAM.ServiceLinkedRoles.html#service-linked-role-permissions`.

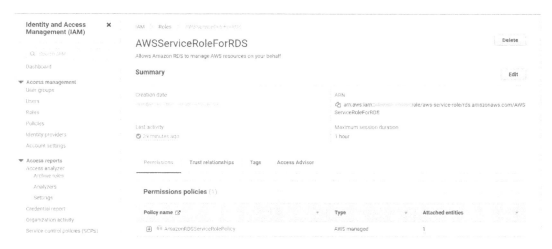

Figure 5.4 – RDS service-linked role to manage AWS Outposts

The advantage of using service-linked roles is that you don't need to manually add permissions. By granting permissions to the Outposts service to make calls to other AWS services on your behalf, only AWS Outposts can assume these roles. You can use trust policies to prevent other IAM entities to use these roles.

Let's look at how AWS Outposts uses service-linked roles to access the AWS resources needed for private connectivity. One role will be created with the name `AWSServiceRoleForOutposts_<OutpostID>`, allowing the creation of network interfaces needed for private connectivity and attaching them to service link endpoint instances.

The `AWSServiceRoleForOutposts_<OutpostID>` service-linked role trusts the `outposts.amazonaws.com` service to assume the role:

Figure 5.5 – Service-linked role trust to allow only the AWS Outposts service to assume the role

The `AWSServiceRoleForOutposts_<OutpostID>` service-linked role has the following policies attached:

- `AWSOutpostsServiceRolePolicy`
- `AWSOutpostsPrivateConnectivityPolicy_<OutpostID>`

Figure 5.6 – AWS Outposts service-linked role attached policies

Let's inspect these policies. `AWSOutpostsServiceRolePolicy` is a service-linked role policy to enable access to AWS resources managed by AWS Outposts and the `AWSOutpostsPrivateConnectivityPolicy_<OutpostID>` policy allows AWS Outposts to manage resources used to establish private connectivity.

Figure 5.7 – Policy details for Outposts service-linked role

Here are couple more examples of service-linked roles trusting only one service to assume the role. In the first example, the `eks.amazonaws.com` EKS service is trusted to assume the following role to manage clusters and perform actions on your behalf in AWS Outposts:

Figure 5.8 – Service-linked role trust to allow only the EKS service to assume the role

This role has four policies attached; two are *AWS-managed policies* and the other two are *customer-managed policies*. To find out more about managed policies and inline policies, check this URL: `https://docs.aws.amazon.com/IAM/latest/UserGuide/access_policies_managed-vs-inline.html`.

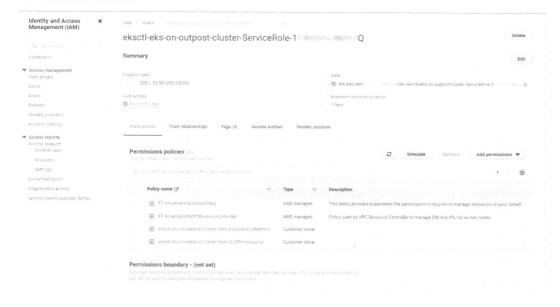

Figure 5.9 – AWS EKS service-linked role attached policies

The customer-managed policies were created to allow EKS to push metric data to Amazon CloudWatch and manage elastic load balancers. Here is the content of the `eksctl-eks-on-outpost-cluster-PolicyCloudWatchMetrics` policy:

```
{
    "Version": "2012-10-17",
    "Statement": [
        {
            "Action": [
                "cloudwatch:PutMetricData"
            ],
            "Resource": "*",
            "Effect": "Allow"
        }
    ]
}
```

Here is the content of the `eksctl-eks-on-outpost-cluster-PolicyELBPermissions` policy:

```
{
    "Version": "2012-10-17",
    "Statement": [
        {
            "Action": [
                "ec2:DescribeAccountAttributes",
                "ec2:DescribeAddresses",
                "ec2:DescribeInternetGateways"
            ],
            "Resource": "*",
            "Effect": "Allow"
        }
    ]
}
```

In the second example, the `ec2.amazonaws.com` EC2 service is trusted to assume the following role to manage clusters and perform actions on your behalf in AWS Outposts:

Figure 5.10 – Service-linked role trust to allow only the EC2 service to assume the role

This role has two AWS-managed policies attached to allow the EKS node to perform actions on Amazon CloudWatch:

Figure 5.11 – AWS EKS node service-linked role attached policies

AWS Outposts makes extensive use of AWS IAM and interoperates with other AWS security services offered in the Region. Combined with AWS CloudTrail, AWS CloudWatch, AWS Organizations, or AWS Control Tower, IAM can provide an enterprise-grade authentication, authorization, and accounting infrastructure to implement your guardrails and controls

The next section will talk about the physical portion of security affecting anyone using Outposts. It may surprise those of you accustomed to never thinking about the security of physical things using AWS because you don't have to care about it whenever you are operating within the confines of an AWS Region.

The physical security of Outposts

After understanding how AWS Outposts integrates seamlessly with the AWS ecosystem of security services and uses its capabilities, let's talk about the physical security of Outposts hardware.

This may be a new topic for some of you, especially for cloud-native professionals. You don't get into these details when you build in the Region. You know that, of course, there is hardware somewhere in AWS data centers but it is AWS's responsibility to keep it secure and running.

When operating within the AWS Region, customers are bound by the AWS Shared Responsibility Model. The purpose of this statement is to clearly demarcate who is responsible for what when you select the AWS platform to run your applications. By operating under this framework, the customer is relieved from any operational burden in regard to facilities supporting the cloud operations.

AWS defines this model by stating *"AWS is responsible for security of the cloud and the Customer is responsible for security in the cloud"* in its Shared Responsibility Model. To summarize, AWS is responsible for security up to the point where the customer is granted access to resources. From that point on, the security of the resource is the responsibility of the customer.

Let us give an example. Suppose a customer creates an EC2 instance with a security key associated. During the time the instance is being provisioned, it is still under AWS's control and it is AWS's responsibility to make sure the OS does not contain any malicious code and no extraneous content is injected during the provisioning process.

By the time the hypervisor boots the OS, the instance is under your control and whatever happens after that is your responsibility. That includes configurations, user data scripts, and cloud-init directives that you specify when you launch the instance.

The same principle applies to other AWS services; the moment you are allowed to control the resource and apply configurations, your responsibility begins. AWS summarizes this framework as per *Figure 5.12*, and you can find more information at this URL: `https://aws.amazon.com/compliance/shared-responsibility-model/`.

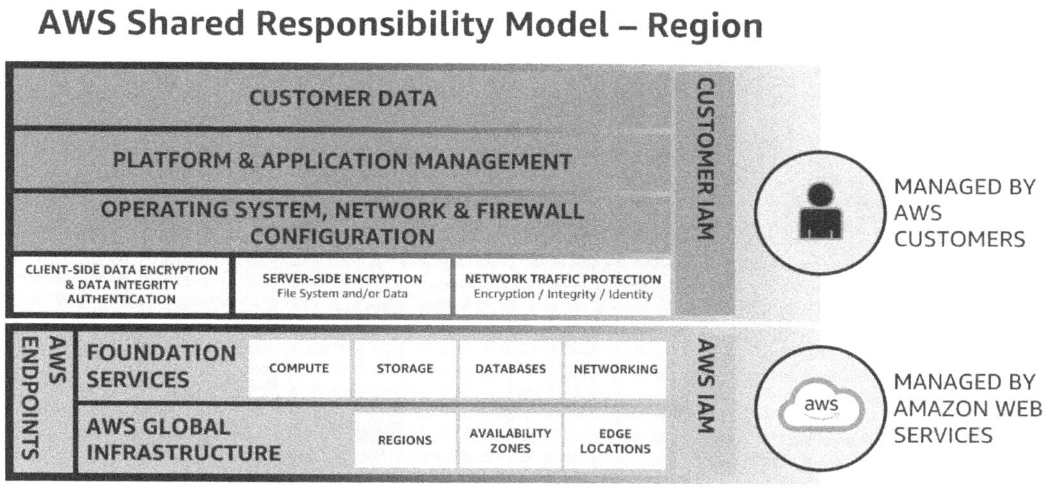

AWS Shared Responsibility Model – Region

Figure 5.12 – AWS Shared Responsibility Model

However, Outposts operates outside AWS boundaries and this framework needs to be adjusted to consider this fact. When we think about the implications in terms of risk and exposure when we deploy an extension of AWS on the customer premises, the initial feeling is that this framework would become extremely complex.

Turns out that AWS engineering did an outstanding job at implementing security features and guardrails to minimize the risks derived from this exposure and allow a thoroughbred AWS piece of equipment, the same used by AWS to build its Regions, to be deployed in your data center. Let us amend the Shared Responsibility Model so it fits the nuances imposed by deploying AWS technology outside its walls.

The result is to include one additional layer in this framework under the responsibility of the customer, represented in *Figure 5.13*. These new duties were detached from the Shared Responsibility Model applied in the Region to account for the data center portion now under your control. Everything else remains unchanged:

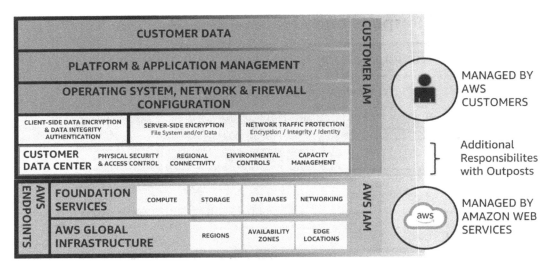

Figure 5.13 – AWS Shared Responsibility Model for AWS Outposts

The adjusted framework is straightforward. You are now in charge of providing and assuring the quality, smooth operation, and continuity of the following aspects:

- **Physical security and access control**: You are responsible for guaranteeing that only authorized personnel can physically reach the rack. Situations arising from unintended physical access of the rack could potentially result in disconnection of network uplinks, disconnection of power, attempts to violate the rack and access its components, and attempts to move or roll the rack from its seating position.

- **Regional connectivity**: AWS Outposts is not meant to operate in disconnected mode; it cannot survive disconnections longer than a couple of hours without any guaranteed limits. AWS Outposts is not a solution that can operate without communication with the parent Region for a long period of time. It is the customer's responsibility to ensure that the link is stable and reliable.

- **Environmental controls**: You are responsible for controlling the temperature, airflow, and humidity within your facility. Also, you must provide a stable location supporting the weight of the rack, seismic anchorage when required, and enough clearance to maneuver the rack and open its doors.

- **Capacity management**: In the AWS Region, you don't need to worry whether you have enough capacity to launch instances, as you have *virtually unlimited* capacity to operate. However, AWS Outposts has capacity limits. You will be informed about these limits and all the tooling necessary to monitor and control resource usage is available to you. AWS Outposts will throw an *insufficient capacity* error if it runs out of resources and it will not automatically fall back to the Region. Additional capacity needs to be added.

But how exactly did AWS engineering decisions help to alleviate the burden imposed on customers due to these changes in the security model? The answer relates to the security features implemented in the rack:

- Built-in tamper detection

- Enclosed rack with a lockable door

- Removable and destroyable hardware security key on each server (NSK)

Effectively, the customer cannot open the rack nor perform any actions on it. Everything is carried out by AWS personnel onsite or remotely. Droplets do not have USB ports or traditional management interfaces. You can see and touch the hardware, but you cannot manipulate or service the rack components.

Are we done with security after having extensive lectures on logical and physical security? Oh no, not yet. Hardware is not perfect; it does fail eventually. When your design is affected by the physical placement of hardware and constrained by several available elements to support the workload, it is also your duty to implement strategies to allow your app to withstand the failure of the underlying infrastructure or resume operations on spare capacity.

In the next section, we will discover how to implement strategies to improve the resiliency of our Outposts and improve our application availability and uptime serving customers.

Outposts resilience

Physical security is particularly important for the continuity of operations. You do not want your system to be disrupted by a power loss caused by someone pulling the power whip. However, problems can also happen to rack components even if they are kept inaccessible inside a cage.

Your criteria to select the rack SKU and your strategy to deploy applications and solutions on Outposts must always consider various levels of failure, external and internal to the rack or problems with services it depends on, such as the control plane in the Region. As Werner Vogels, vice president and CTO of Amazon, says:

Everything fails all the time.

Each time we introduce a technical term, I consider it a best practice to establish a common ground about what it means. Once again, I will use the NIST definition of *resilience*. If you head over to the URL https://csrc.nist.gov/glossary/term/resilience, you will actually find a dozen definitions. I will use the simplest of all of them as a basis for discussion:

The ability to maintain required capability in the face of adversity.

This concept requires the matter to be expressed in degrees of resiliency. With this nuance, the more rapidly and effectively a system protects its capabilities, the more resilient it is. Any system that is 100 percent resilient to all adversities is unlikely to be practically or financially feasible.

For this reason, critical systems normally incorporate multiple resilience controls. These controls can support detection, response, or recovery. When defining requirements, you must clearly delineate the levels of resilience to be expected from all multiple subsystems that make up one complex system.

The design of the rack already implements various levels of redundancy and survivability. **Outpost Networking Devices (ONDs)** are redundant, power components operate in pairs, and Droplets have redundant connections to ONDs. The probability of simultaneous failure of these components is exceptionally low, which translates into an extremely high degree of resilience.

If you lose a component, the redundant component can keep the system operating without significant impact, aside from losing the ability to tolerate the fault of the component, which is now the single point of failure and can render the entire system useless if it fails. AWS will detect the failure of the component and respond by triggering its replacement. The system will not cease functionality, and once the faulty component is replaced, the system will automatically recover to its full potential.

But what about the Droplets used to deliver compute and storage capacity (EBS)? The high availability and fault tolerance of the EBS service's physical subsystem is the responsibility of AWS; however, the high availability and fault tolerance of applications and systems that use compute elements and AWS services running on Outposts is your responsibility.

Designing and implementing a high availability and fault tolerance strategy in regard to application resilience is your responsibility. You must define strategies and select tools to account for the faulty elements or conditions, such as the following:

- Break in regional communication
- Break in on-premise network communication
- Failure of EC2 instances
- Insufficient storage capacity
- Insufficient compute capacity
- Failure of instance store modules (NVMe)
- Failure of one entire compute Droplet
- Failure of an Availability Zone at the parent Region
- Failure of an entire Region

Let us once again set a common ground for our discussion and check the NIST definitions for *high availability* and *fault tolerance*.

The NIST definition for high availability is as follows (https://csrc.nist.gov/glossary/term/high_availability):

A failover feature to ensure availability during device or component interruptions.

The NISRT definition for fault tolerance is as follows (`https://csrc.nist.gov/glossary/term/fault_tolerance`):

A property of a system that allows proper operation even if components fail.

There are two analogies that have helped to illustrate these concepts over time. This is so important because most people mix up the two and specify that a system must be highly available, but in reality, they want fault-tolerant systems.

For the first example, let's think of a car. You require that the rolling subsystem is composed of tires and rims to be resilient. The proposed solution is to add a spare tire to replace one working tire if it gets flat or if you have a damaged rim:

- *Is it highly available?* Yes, the spare tire and tools included with the car ensure that the subsystem can be brought to function after replacing the flat tire, therefore you do have a failover feature. Consider that the propulsion subsystem represented by the car engine is not highly available; you don't have a failover option if it breaks.

- *Is it fault tolerant?* No, the car will not operate properly with a flat tire. You will need to stop, replace the tire, and resume your trip.

- *Is it resilient?* To an average degree, yes. In the event of a malfunction, the system is not able to maintain its capability and the car needs to stop. But it can be fixed quickly because it has built-in features to be put back to work. Imagine if you do not have a spare; a flat tire will put a halt to your trip for a great deal of time.

For the second example, let us think of a plane. You require that the propulsion subsystem composed of jet engines be resilient. The proposed solution is to use multiple jet engines to propel the aircraft in an n+1 arrangement. An aircraft with four engines can fly normally using only three:

- *Is it highly available?* Yes, one faulty engine can be shut down and the flight will continue normally, therefore you do have a failover feature. Consider that the aircraft needs three jet engines to be propelled and one extra was added to be a hot spare.

- *Is it fault tolerant?* Yes, the aircraft will operate properly without one jet engine; you do not need to land as quickly as possible. To implement even higher degrees of resilience, the aircraft could be designed to fly for some time with just two jet engines.

- *Is it resilient?* To an extremely high degree, yes. In the event of a malfunction, the system will be able to maintain its capability. It does not need immediate fixing. Even in the case of a serious issue, such as two faulty engines, the aircraft can continue to fly long enough to land safely.

How does it affect our architectural design of compute elements? You need to define resilience requirements for applications and specify whether they need high availability or fault tolerance. AWS does not have a feature similar to VMware vSphere Fault Tolerance. It is your duty to use the app's built-in features and tools or third-party apps to implement resilience.

Levels of resilience may dictate your system may need spare Droplets in one rack, Droplets deployed to multiple racks in a data center, or racks associated with different Availability Zones placed at different data centers over metro areas.

There is an extensive AWS whitepaper about high availability that is a must-read, at this URL: `https://docs.aws.amazon.com/whitepapers/latest/aws-outposts-high-availability-design/aws-outposts-high-availability-design.html`.

Here are some important points to always keep in your mind to design resilient Outpost architectures:

- An EC2 instance of a given type and size can only be started in a Droplet that supports that exact type and size. Some examples are as follows:

  - A compute-optimized instance (C5) can only be started in a Droplet that supports that family, never on an M5 or R5 Droplet. Droplets only support one instance type; there are no hybrid Droplets.

  - A c5.2xlarge instance can only be started in a Droplet with an available c5.2xlarge slot. It will not start in a different slot, not even a larger slot such as a c5.4xlarge. You can change the instance size to make it fit into an available slot in a C5 Droplet.

- You can segregate instances using placement groups. More information can be found at this URL: `https://docs.aws.amazon.com/AWSEC2/latest/UserGuide/placement-groups.html#placement-groups-spread`.

- Thoroughly understand the slotting process as described in the high-availability whitepaper. Re-slotting a Droplet requires all running instances to be evicted from that Droplet and that means stopping and starting it elsewhere.

Whenever you explore an application's built-in features for continuity, do not forget that you may not need exact duplicates of individual components. This is the easiest strategy, but it can lead to having custom rack SKUs comprised of many distinct Droplet families. One good practice that is not immediately apparent is that homogeneous racks comprised solely of one or two Droplet families are easier to maintain and implement high availability.

In a distributed system comprised of several instances of different sizes, you can possibly evaluate whether the running components can be covered by larger instances of other families. You just need to change the EC2 instance type configuration before starting it. M5 instances could be a good backup for R5 instances, for example.

Another popular strategy is to have the required production system capacity to be covered by the QA or test system. If there is a failure that affects the productive system, the QA or test system is shut down so the production system can take those slots and resume operations.

Now that we have end-to-end security implemented, it is time to discover how much we can trust the hardware and software provided by AWS. In the next section, we will talk about the certifications awarded to Outposts and how it ranks against compliance standards.

Compliance and certifications

There are several aspects related to security that fall under your responsibility; a portion of an AWS data center is under your possession, after all. Moreover, AWS is trusting architectural decisions to your technical abilities and criteria, therefore it is reasonable that accountability for those decisions falls within your scope.

If you are looking for assurances about the quality of the hardware and software, the answer can be found in the certifications and attestations awarded to AWS Outposts. AWS has an extensible and complete certification portal. It is really worth visiting this page: `https://aws.amazon.com/compliance/`.

AWS proudly claims to have the broadest set of compliance offerings today, supporting more security standards and certifications than anyone else. These accreditations include PCI-DSS, HIPAA/HITECH, FedRAMP, GDPR, FIPS 140-2, and NIST 800-171.

The **General Data Protection Regulation (GDPR)** is a singular topic, affecting businesses operating in the **European Union (EU)**. If you are wondering whether the AWS Shared Responsibility Model is applicable in the context of GDPR, the short answer is yes, and the more elaborate answer can be found in an AWS blog at this URL: `https://aws.amazon.com/blogs/security/the-aws-shared-responsibility-model-and-gdpr/`.

It shouldn't be surprising to learn that the AWS Outposts rack will go through a separate evaluation for certifications, and existing certifications will not apply. AWS Outposts has a separate list of services in scope for the specific compliance or assurance program.

The AWS Outposts rack is HIPAA eligible, PCI, SOC, ISMAP, IRAP, and FINMA compliant, ISO, CSA STAR, and HITRUST certified, and GxP compatible. You can find a list of all AWS services that are covered at this URL: `https://aws.amazon.com/compliance/services-in-scope/`.

AWS Outposts obtained FedRAMP authorization status with the exclusion of the hardware components providing the service. Customers can decide to take the risk or conduct an assessment to review the hardware components of AWS Outposts to process FedRAMP workloads because they are responsible for physical and environmental controls for AWS Outposts.

Lastly, if you want to access the AWS compliance reports, you can do so by checking AWS Artifact, where you have on-demand access to AWS's security and compliance reports and select online agreements. AWS Artifact central can be found at this URL: `https://aws.amazon.com/artifact/`. But the reports can be found directly at this AWS Management Console URL: `https://console.aws.amazon.com/artifact`.

We are done with covering the security aspects within our scope for AWS Outposts. Security is so important that you can find several books dedicated to covering only security for AWS. You have an AWS certification accreditation dedicated to security. You will never be done with security; take it very seriously and you will never regret your decision.

Summary

In this chapter, we have covered most security aspects pertaining to Outposts. You can never have enough information in regard to security; the topic is so vast and complex. I encourage people to specialize in security. It is frequently a neglected area of knowledge because, well, when you ask people about it, the most common answer is, "*I think we are secure.*"

The intention here was not to make you a CISO, but to enable you to be an authoritative reference and act as an advisor in regard to embedding security practices and principles when selecting, implementing, and operating your Outposts. If we have succeeded at this, your response would probably be, "*I know how secure we are and the risks we assumed.*"

You should now know how to implement security for your data, your entities, and your facilities. You know how to design resilient services and apps running on Outposts and how to handle inquiries about compliance and certifications awarded to Outposts.

We are approaching the conclusion of our journey. Take a moment to reflect on all you have learned so far about Outposts and this realm called *hybrid edge*, and if you feel proud of yourself, it is more than justified.

There is still fun lying ahead, and in the next chapter, we will talk about monitoring your Outposts because you can't manage what you can't measure and you must trust but verify. This is a play on some famous quotes, but hopefully, now I have you hooked on the subject. Let's go!

6
Monitoring Outposts

You are now confident and enabled to implement security and compliance around your hybrid architecture. Internal security threats originating within an organization from a current or former employee, a contractor, or a business associate are often underestimated.

No company is static; change is the only constant. It is not enough to create a process and train people to do the right thing; you also have to make sure people are not engaging in any wrongdoing behind your back. Moreover, it is not only people that fail; hardware also fails. Systems are designed to operate within certain limits, but what happens when the upper or lower limits are breached?

The obvious answer is that you must act in response to these events, but how can you become aware that something is happening within your walls? The answer is that you must monitor and be alerted about these events, which is what this chapter is about. It will cover the following:

- Monitoring the capabilities of Amazon CloudWatch
- Alarms, dashboards, and custom queries with CloudWatch
- The logging capabilities of AWS CloudTrail
- Querying CloudTrail logs with Amazon Athena

Monitoring with CloudWatch

After thoroughly understanding the security aspects of AWS Outposts, we will now discover how we can monitor our Outpost's capacity and improve the user experience. We collect metrics to identify and resolve issues before they impact our customers. Remember that the capacity of Outposts is finite, and it is your responsibility to manage the resources available to you.

You don't have access to AWS Outposts telemetry; this is the responsibility of AWS. Hardware degradation or failure should be detected by AWS teams, and the replacement process will be triggered.

You must embrace failure and design architecture to be resilient and withstand failures at various levels, potentially up to the unavailability of an entire AWS Region. However, failover procedures or election processes among nodes of a distributed application must be triggered by some sort of change in your system.

The AWS go-to service to collect metrics, monitor your systems, create alarms, and trigger automated responses to events is Amazon CloudWatch. CloudWatch goes beyond the basics and can analyze your data and generate insights from applications. CloudWatch provides a rich set of graphics and visualization options so that you can build very complex dashboards.

Amazon CloudWatch Events delivers a continuous stream of system events resultant from changes in AWS service resources; this is called a **trail**. You can set up rules that match these events and take actions, invoking a certain target AWS service when it finds a match.

For example, it can send SNS notifications or take a corrective measure, executing a Lambda function. A comprehensive list of supported targets can be found at this URL: `https://docs.aws.amazon.com/AmazonCloudWatch/latest/events/WhatIsCloudWatchEvents.html`. Amazon recommends evaluating Amazon EventBridge, a serverless event bus that has the same underlying service and API as CloudWatch Events but with more features. Here is a reference URL to begin discovering this service: `https://docs.aws.amazon.com/eventbridge/latest/userguide/eb-what-is.html`.

To react to metrics breaching upper or lower thresholds for a certain duration, you can use CloudWatch Alarms, to which you can also have multiple actions attached. Once an alarm is triggered, these actions are invoked to perform tasks defined for each action.

These actions are limited when compared to the capabilities of CloudWatch events. You can find more information at this URL: `https://docs.aws.amazon.com/AmazonCloudWatch/latest/monitoring/AlarmThatSendsEmail.html`.

CloudWatch in action – alarms

Let's have a look at CloudWatch in action. Head on to the CloudWatch service landing page, and you will arrive directly at the **Overview** page, shown in the following screenshot; at this particular time, there are no active alarms, no default dashboards, or Application Insights defined.

For the purposes of this demonstration, the CloudWatch screenshots shown here were taken from an already configured Outpost in a Lab. When you do this for the first time, your CloudWatch console page may look radically different, but the concepts are the same.

Select **All alarms** on the left pane, and we can check all defined alarms.

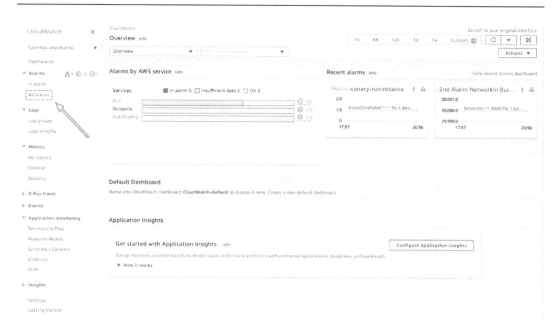

Figure 6.1 – Amazon CloudWatch landing page – Overview

On the **Alarms** page, let's examine the configuration of one given alarm. You can do this by clicking on the alarm name.

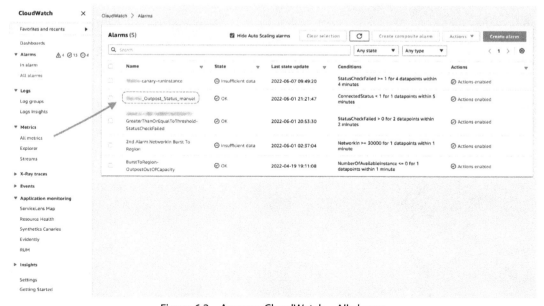

Figure 6.2 – Amazon CloudWatch – All alarms

You will be taken to the details page of the selected alarm. At the bottom section, we can identify four distinct tabs; let's peek into them: **Details**, **Actions**, and **History**. On the **Details** tab, you will have a summary of definitions for the selected *alarm*. Look at the defined **Threshold**, **Statistic**, and **Datapoints to alarm** configurations.

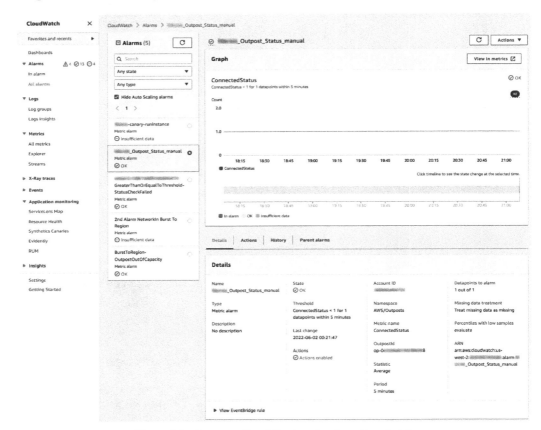

Figure 6.3 – Amazon CloudWatch – the alarm Details tab

The **Actions** tab will show the types of tasks defined to be executed once the *alarm* is triggered.

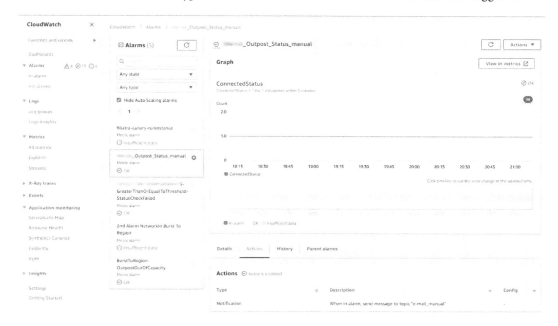

Figure 6.4 – Amazon CloudWatch – the alarm Actions tab

As for the **History** tab, it shows a list of dates and times when the defined *alarm* transitioned its state.

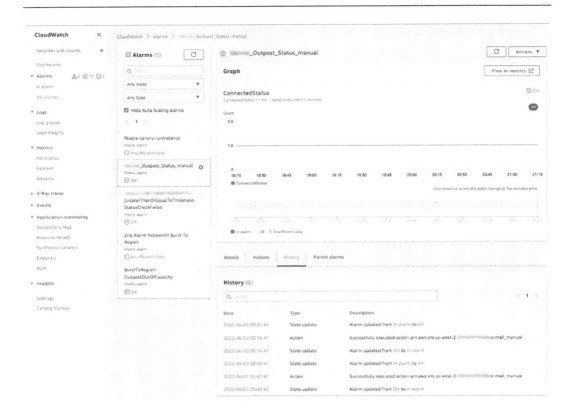

Figure 6.5 – Amazon CloudWatch – the alarm History tab

Now, let's look at CloudWatch's graphic capabilities.

CloudWatch in action – dashboards

On the left pane, click **Dashboards**, and you will be taken to a page with two distinct tabs at the bottom. By default, you will land on the **Custom dashboards** tab; the other tab is **Automatic dashboards**. On the list of custom dashboards, you can access the graphical portion of the page by clicking on the custom dashboard name.

Figure 6.6 – Amazon CloudWatch – Custom dashboards

You will be taken to the widgets page. As seen in the following screenshot, four widgets were added to this view, each displaying a particular metric. The graphic represents the measured dimension values at specific intervals, plotted over the defined period of time selected; in this view, we chose **Custom (3M)** to visualize data points recorded over a period of 3 months.

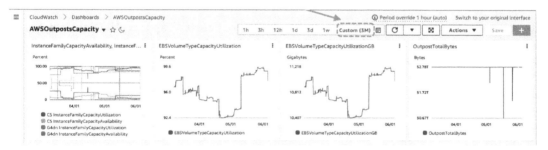

Figure 6.7 – Amazon CloudWatch – the widgets of one dashboard

For a detailed view of any widget, you just need to maximize it; the maximize icon is shown when you hover with the mouse over any widget. The expanded view gives better visualization; the following are the expanded views of three of the widgets shown in *Figure 6.7*.

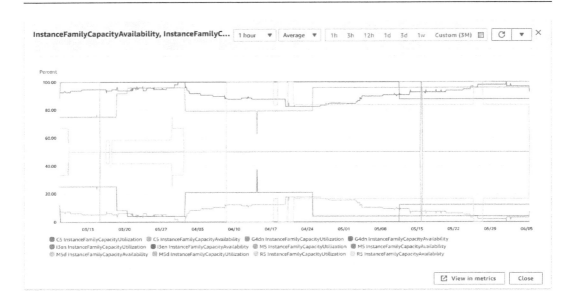

Figure 6.8 – Amazon CloudWatch – a maximized view of Instance Family Capacity widgets

The preceding widget shows two *metrics*, **InstanceFamilyCapacityUtilization** and **InstanceFamilyCapacityAvailability** for **C5**, **I3en**, **G4dn**, **M5**, **M5d**, and **R5**, for the **InstanceFamily** dimension using the **Average** statistic.

Figure 6.9 – Amazon CloudWatch – a maximized view of the EBSVolumeTypeCapacityUtilization widget

The preceding widget shows the **EBSVolumeTypeCapacityUtilization** metric for the **OutpostId** dimension, using the **Average** statistic.

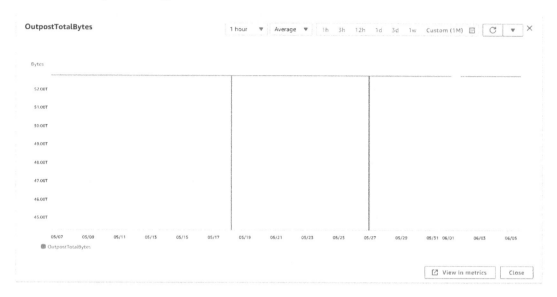

Figure 6.10 – Amazon CloudWatch – a maximized view of the S3 OutpostTotalBytes widget

The preceding widget shows the **OutpostTotalBytes** metric for the **OutpostId** dimension, using the **Average** statistic.

Other metrics and widgets are available; you can find an extensive list at this URL: `https://docs.aws.amazon.com/outposts/latest/userguide/outposts-cloudwatch-metrics.html`. Next, I will show the **ConnectedStatus** and **CapacityExceptions** *metrics* for the **OutpostId** dimension.

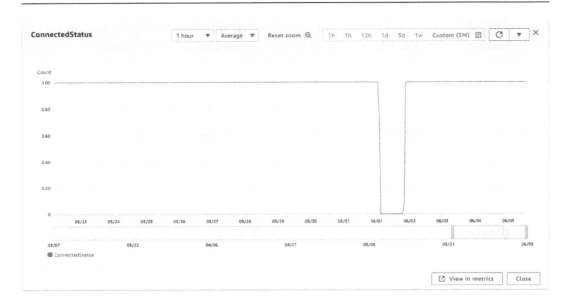

Figure 6.11 – Amazon CloudWatch – a maximized view of the ConnectedStatus widget

The preceding widget shows the **ConnectedStatus** metric for the **OutpostId** dimension, using the **Average** statistic.

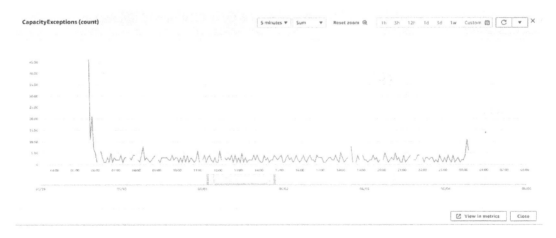

Figure 6.12 – Amazon CloudWatch – a maximized view of the CapacityExceptions widget

The preceding widget shows the **CapacityExceptions** metric for the **OutpostId** dimension, using the **Sum** statistic. For this graphic, the interval is 1 week (**1w**).

CloudWatch visualizations on the AWS Outposts service page

The Outposts service landing page also has visualizations to help monitor your AWS Outposts. Once you get to the page, just click one Outpost ID, and you will be taken to the Outposts' **Summary** page. There are eight tabs available in the bottom section.

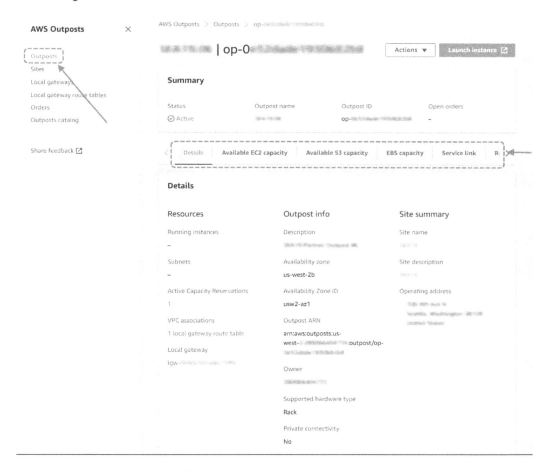

Figure 6.13 – AWS Outposts – the Summary page for one OutpostID

The **Available EC2 capacity** tab has pre-built visualizations for three metrics: **CapacityExceptions (count)**, **CapacityAvailability (%)**, and **CapacityUtilization (%)**. If you click this tab, you will get an expanded view, similar to the one shown in *Figure 6.14*. For this particular Outpost, the first visualization shows the total number of capacity exceptions for one specific instance type (**C5**) during one week (**1w**).

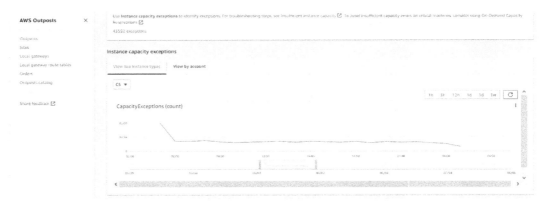

Figure 6.14 – AWS Outposts – capacity exceptions for one instance type

Scrolling down the page, you will find the visualization of **CapacityAvailability (%)**. *Figure 6.15* shows the screenshot for this particular Outpost; by default, it is displaying the percentage of available instances for one specific instance type (**C5**), spanning 4 weeks (**4w**).

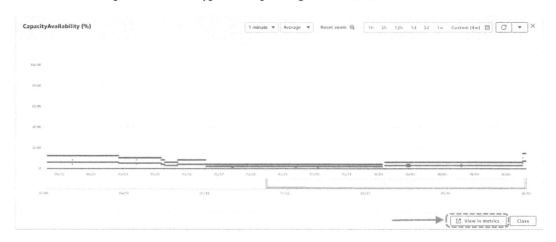

Figure 6.15 – AWS Outposts – capacity availability for one instance type

If you click the **View in metrics** button, you will be taken to the CloudWatch service page where you can change several parameters and configurations.

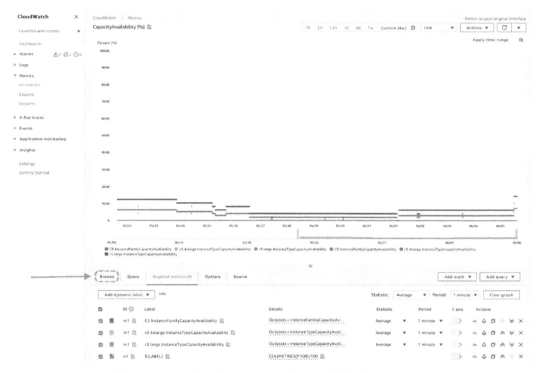

Figure 6.16 – Amazon CloudWatch – capacity availability for one instance type

If you click the **Browse** tab at the bottom section, it will display all metrics that your IAM identity was granted permission to access. More information about granting IAM permission to read metrics can be found at this URL: `https://docs.aws.amazon.com/AmazonCloudWatch/latest/monitoring/iam-identity-based-access-control-cw.html`.

In this particular case, we have metrics for Outposts and S3 on Outposts. More information about CloudWatch metrics available for Amazon S3 on Outposts can be found at this URL: `https://docs.aws.amazon.com/AmazonS3/latest/userguide/metrics-dimensions.html#s3-outposts-cloudwatch-metrics`.

Figure 6.17 – Amazon CloudWatch – browsing all metrics available

If you click on **Outposts**, you will have the metrics grouped by one or more *dimensions*.

Figure 6.18 – Amazon CloudWatch – Outpost metrics grouped by dimension

Click on **By Outpost, Instance Type, and Account** for a detailed view.

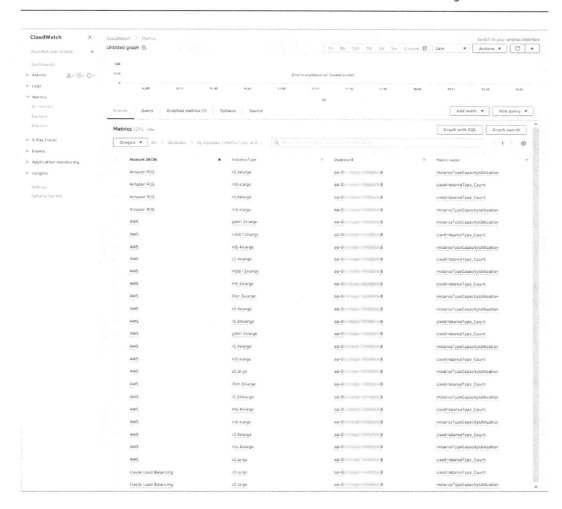

Figure 6.19 – Amazon CloudWatch – Outpost metrics grouped by three dimensions

If you click on **S3Outposts** in *Figure 6.17*, you will have a similar view of the metrics, grouped by one or more *dimensions*.

Figure 6.20 – Amazon CloudWatch – S3 Outpost metrics grouped by dimension

If you click the **Query** tab at the bottom section, you will be taken to **Metrics Insights - query builder**. More information about CloudWatch Logs Insights can be found at this URL: `https://docs. aws.amazon.com/AmazonCloudWatch/latest/logs/AnalyzingLogData.html`.

Figure 6.21 – Amazon CloudWatch – Metrics Insights

For this example, let's click the **Editor** button, where we can write our query using the familiar SQL syntax and structure. A tutorial explaining how to run and modify a query can be found at this URL: `https://docs.aws.amazon.com/AmazonCloudWatch/latest/logs/ CWL_AnalyzeLogData_RunSampleQuery.html`.

The following SQL sentence will display the `Count` aggregation function of the `UsedInstanceType_ Count` metric from the `AWS/Outposts` namespace, grouped by `InstanceType`:

```
SELECT COUNT(UsedInstanceType_Count) FROM "AWS/Outposts" GROUP
BY InstanceType
```

The resulting visualization is as follows. I renamed the resulting query **OP_UsedCapacity**.

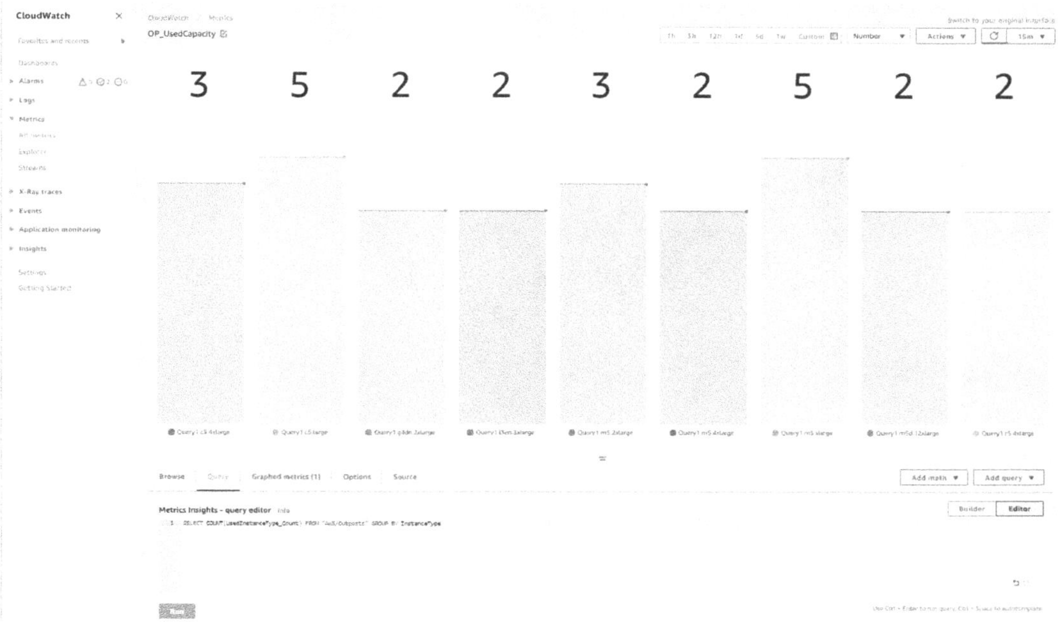

Figure 6.22 – Amazon CloudWatch – the OP_UsedCapacity custom query

Let's run one more query. The following SQL sentence will display the Count aggregation function of the AvailableInstanceType_Count metric from the AWS/Outposts namespace, grouped by InstanceType:

```
SELECT COUNT(AvailableInstanceType_Count) FROM "AWS/Outposts"
GROUP BY InstanceType
```

The resulting visualization is shown as follows. I renamed the resulting query **OP_AvailableCapacity**.

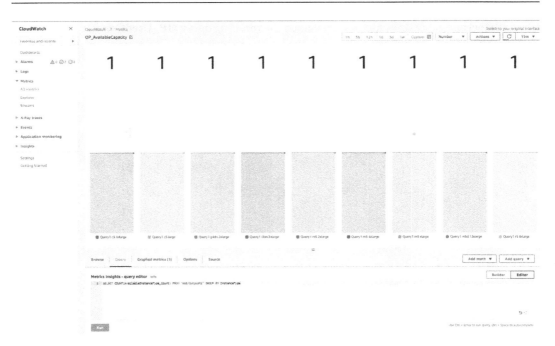

Figure 6.23 – Amazon CloudWatch – the OP_AvailableCapacity custom query

For our last query, the following SQL sentence will display the AVG aggregation function of the InstanceFamilyCapacityAvailability metric from the AWS/Outposts namespace, grouped by InstanceFamily:

```
SELECT AVG(InstanceFamilyCapacityAvailability) FROM "AWS/
Outposts" GROUP BY InstanceFamily
```

The resulting visualization is shown as follows. I renamed the resulting query **OP_AverageAvailableCapacityPerInstanceFamily**.

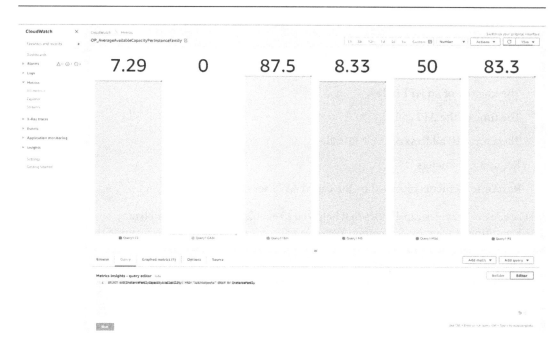

Figure 6.24 – Amazon CloudWatch – the OP_AverageAvailableCapacityPerInstanceFamily custom query

As per the resulting query, the higher the number, the more capacity we have available on average for the defined period (**3h**) in this case.

We are done with our exploration of CloudWatch, but in full honesty, there is much more to learn about it. Amazon CloudWatch is a critical part of the AWS cloud nervous system; any design that needs automation and orchestration will have Amazon CloudWatch playing a critical role.

Take a look at the Amazon CloudWatch documentation center at this URL: `https://docs.aws.amazon.com/cloudwatch/index.html`. You will find more than a dozen URLs for references and documentation about CloudWatch various capabilities. Don't forget Amazon EventBridge, the service designed to help you build event-driven architectures that are distributed and loosely coupled.

Logging with CloudTrail

CloudWatch is a metric gatherer and orchestration service. Metrics can be collected by the service components embedded in other services or sent by agents installed inside operating systems that can run anywhere, including your own premises.

However, CloudWatch is unable to track API calls invoked by AWS identities. The service designed to perform this task is AWS CloudTrail, which helps you record actions taken by users, roles, or other AWS services as events in CloudTrail.

These events can be sent as log files to an Amazon S3 bucket for storage and further analysis of who is doing what and when. An extensive record of data related to the API activity is provided, including the following:

- The identity of an API caller
- The time of the API call
- The source IP address of the API caller
- Request parameters
- Response elements returned by the called AWS service

AWS CloudTrail provides capabilities that help you to enable governance, compliance, and operational and risk auditing of your AWS account. AWS CloudTrail is responsible for performing the *accounting* piece of our **Authentication, Authorization, and Accounting (AAA)** infrastructure.

AWS Outposts has a set of API calls that are captured by CloudTrail. There is an example of a CloudTrail log entry and more information at this URL: `https://docs.aws.amazon.com/outposts/latest/userguide/logging-using-cloudtrail.html#outposts-info-in-cloudtrail`. However, there is a very interesting combination of tools to perform trail analysis – CloudTrail sending logs to an S3 Bucket that can be analyzed by Amazon Athena.

The first step is to configure the *trail* to send logs to an Amazon S3 Bucket. Let's see how this looks in the AWS Management Console. You can find more information about receiving and storing CloudTrail logs at this URL: `https://docs.aws.amazon.com/AmazonS3/latest/userguide/cloudtrail-logging.html`.

First, head over to the AWS CloudTrail service landing page. You will arrive at the **Dashboard** view. Click on the **Trails** menu on the left pane.

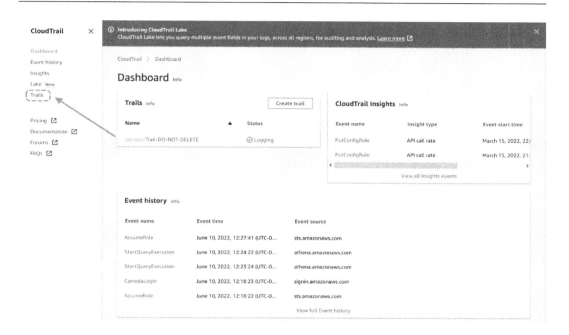

Figure 6.25 – The AWS CloudTrail service landing page

There is a trail already created; let's examine its configurations. If you want to create a new one, you can just click **Create trail**.

Figure 6.26 – AWS CloudTrail Trails page

For this exercise, let's just click the *trail* name and see how it was defined.

Figure 6.27 – AWS CloudTrail – the trail's General details page

Trail log location represents the S3 bucket to be used as the log store repository. If you click that entry, AWS Management Console will take you to the S3 service page, directly into the defined bucket prefixes.

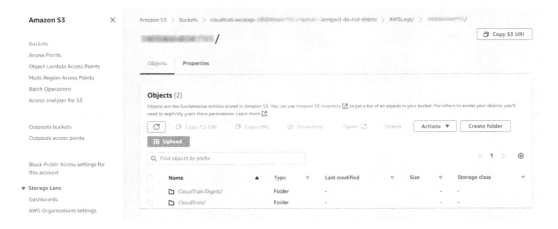

Figure 6.28 – An Amazon S3 logging bucket for one trail

The S3 bucket policy must be configured to allow the CloudTrail service to perform actions on the bucket. You can find more information about how to configure this policy at the following URL: https://docs.aws.amazon.com/awscloudtrail/latest/userguide/create-s3-bucket-policy-for-cloudtrail.html.

Here is an example policy configured for the preceding bucket:

```
{
    "Version": "2012-10-17",
    "Statement": [
        {
            "Sid": "AWSCloudTrailAclCheck20150319",
            "Effect": "Allow",
            "Principal": {
                "Service": "cloudtrail.amazonaws.com"
            },
            "Action": "s3:GetBucketAcl",
            "Resource": "arn:aws:s3:::cloudtrail-awslogs-
123456789012-abcdefghi-jklmnop-do-not-delete"
        },
        {
            "Sid": "AWSCloudTrailWrite20150319",
            "Effect": "Allow",
            "Principal": {
                "Service": "cloudtrail.amazonaws.com"
            },
            "Action": "s3:PutObject",
            "Resource": "arn:aws:s3:::cloudtrail-
awslogs-123456789012-abcdefghi-jklmnop-do-not-delete/
AWSLogs/123456789012/*",
            "Condition": {
                "StringEquals": {
                    "s3:x-amz-acl": "bucket-owner-full-control"
                }
            }
        },
        {
            "Sid": "AWSCloudTrailHTTPSOnly20180329",
```

```
            "Effect": "Deny",
            "Principal": {
                "Service": "cloudtrail.amazonaws.com"
            },
            "Action": "s3:*",
            "Resource": [
                "arn:aws:s3:::cloudtrail-awslogs-123456789012-
abcdefghi-jklmnop-do-not-delete/AWSLogs/123456789012/*",
                "arn:aws:s3:::cloudtrail-awslogs-123456789012-
abcdefghi-jklmnop-do-not-delete"
            ],
            "Condition": {
                "Bool": {
                    "aws:SecureTransport": "false"
                }
            }
        }
    ]
}
```

Now that you have CloudTrail configured to send the logs to one S3 bucket, the next step is querying these logs using Amazon Athena.

Query CloudTrail logs with Amazon Athena

This service allows you to perform interactive queries using SQL statements against data sources, one of which is Amazon S3. You can find more information about the supported data sources at this URL: https://docs.aws.amazon.com/athena/latest/ug/work-with-data-stores.html.

There is an excellent guide on how to query AWS service logs using Amazon Athena. You can find documentation showing you how to configure Amazon Athena to consume logs for several services at this URL: https://docs.aws.amazon.com/athena/latest/ug/querying-AWS-service-logs.html.

Let's now use the AWS Management Console to access Amazon Athena. The AWS Management Console link for the Amazon Athena service takes you to the landing page, where you can clearly identify the **Explore the query editor** button.

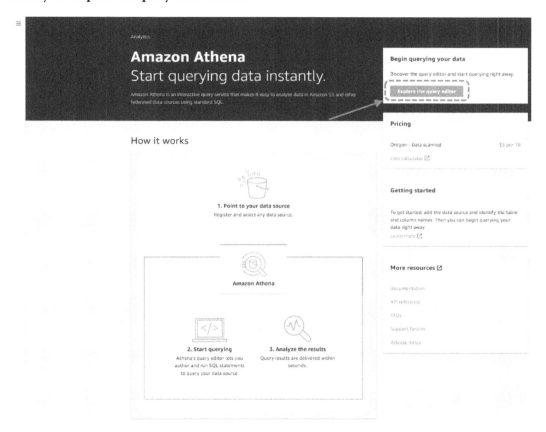

Figure 6.29 – The Amazon Athena landing page

Once you click the button, you will be directed to the **Query editor** interface. Before you execute any queries, you have to define a query result location.

Figure 6.30 – Amazon Athena – the Query editor page

Click **View settings**, and you will be taken to the **Query result and encryption settings** page. Click the **Manage** button to configure the location.

Figure 6.31 – The Query editor settings page

Fill in the required information and click **Save**.

Manage settings

Manage settings

Query result location and encryption

Location of query result
Enter an S3 prefix in the current region where the query result will be saved as an object.

s3://bucket/prefix/object/

View ↗   Browse S3

Expected bucket owner
Specify the AWS account ID that you expect to be the owner of your query results output location bucket.

Enter AWS account ID

Encrypt query results
☐ Enable

☐ **Assign bucket owner full control over query results**
Enabling this option grants the owner of the S3 query results bucket full control over the query results. This means that if your query result location is owned by another account, you grant full control over your query results to the other account.

Cancel   Save

Figure 6.32 – Query editor – the Manage settings page

In our example, the following parameters were used:

- **Location of query result**: s3://aws-athena-query-results-123456789012-us-east-1/QueryResults/

- **Expected bucket owner**: 123456789012

Figure 6.33 – Query editor – the Manage settings parameters

Now, click the **Editor** tab to move back to the **Data** interface, as shown in *Figure 6.33*. You are ready to follow the tutorial to create an Athena table for CloudTrail logs, found at this URL: `https://docs.aws.amazon.com/athena/latest/ug/cloudtrail-logs.html`. For our example, the end result is shown in the following screenshot.

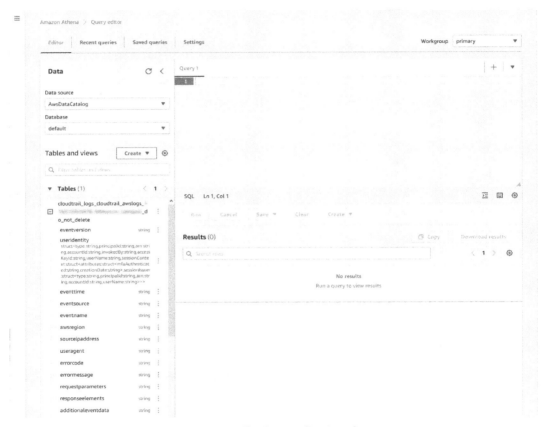

Figure 6.34 – The Query editor interface

Now, the fun begins. In the **Query 1** pane, type the following SQL statement:

```
SELECT
    useridentity.type, eventname, awsregion,
    sourceipaddress, requestparameters,
    useridentity.principalid, useridentity.arn,
    useridentity.accountid, useridentity.invokedby,
    useridentity.accesskeyid, useridentity.username,
    useridentity.sessioncontext, eventsource,
    useragent, errorcode, errormessage,
    responseelements, additionaleventdata, requestid,
    eventid, resources, eventtype, apiversion, readonly,
    recipientaccountid, serviceeventdetails,
    sharedeventid, vpcendpointid
FROM
    cloudtrail_logs_cloudtrail_awslogs_123456789012_abcdef
g_hijklmnop_do_not_delete
WHERE
    CAST(eventsource AS VARCHAR) = 'outposts.amazonaws.com'
GROUP BY
    awsregion, eventname, useridentity.type,
    sourceipaddress, requestparameters,
    useridentity.principalid, useridentity.arn,
    useridentity.accountid, useridentity.invokedby,
    useridentity.accesskeyid, useridentity.username,
    useridentity.sessioncontext, eventsource, useragent,
    errorcode, errormessage, responseelements,
    additionaleventdata, requestid, eventid, resources,
    eventtype, apiversion, readonly, recipientaccountid,
    serviceeventdetails, sharedeventid, vpcendpointid
ORDER BY awsregion, useridentity.type
LIMIT 10000
```

Here is my result; I renamed my query **QueryCloudTrailLogs**:

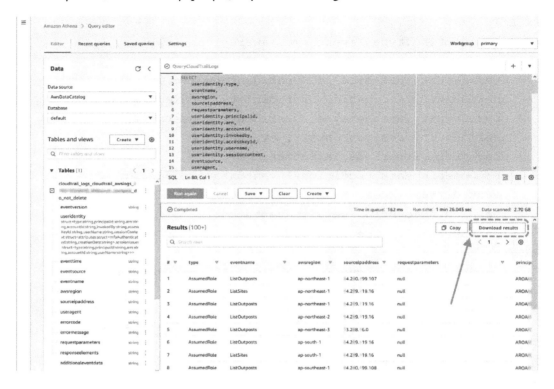

Figure 6.35 – The Query editor results

If you click the **Download results** button, a **Comma-Separated Values** (**CSV**) export file will be downloaded, which you can save anywhere you like.

If you open it with Microsoft Excel, you can perform transformations and filters. I will show some of the distinct results for the userindentity.type, eventname, and requestparameters fields.

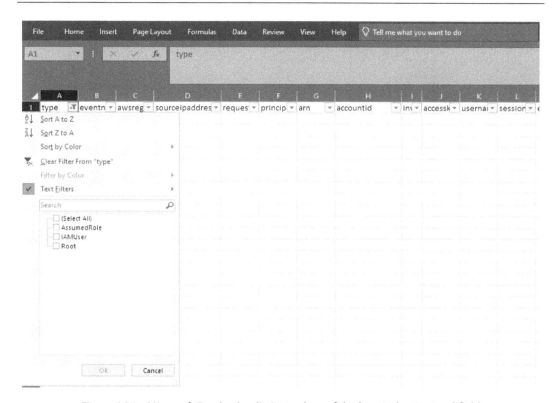

Figure 6.36 – Microsoft Excel – the distinct values of the [userindentity.type] field

The **A** column in Excel shows the returned results of the `userindentity.type` field. You can see three distinct values, including the `Root` account.

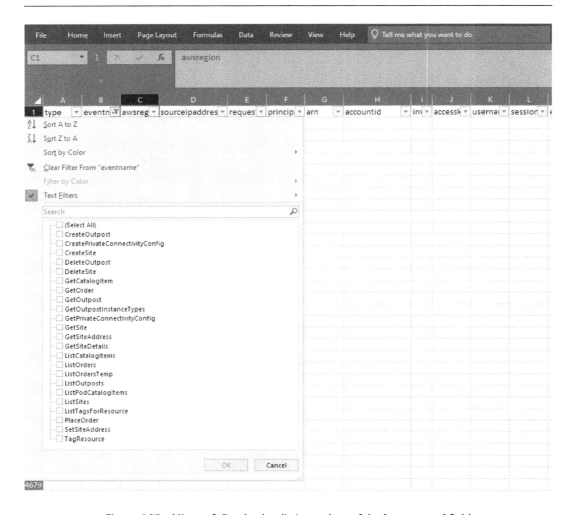

Figure 6.37 – Microsoft Excel – the distinct values of the [eventname] field

The **B** column in Excel shows the returned results of the eventname field. You can see 23 distinct values.

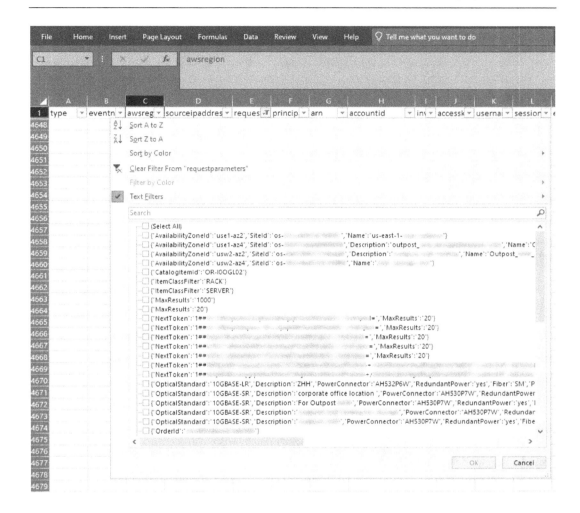

Figure 6.38 – Microsoft Excel – the distinct values of the [requestparameters] field

The **E** column in Excel shows the returned results of the `requestparameters` field. This field shows the parameters used for the API call outlined by `eventname`.

This last exercise brings us to the end of this chapter; I hope you have had fun playing with services and tools to monitor your AWS Outposts. Monitoring is a very important activity in order to collect data, turn it into information, and make management decisions.

Other important combinations of capabilities worth mentioning are as follows:

- The ability to query VPC flow logs with Amazon Athena; more information is available at this URL: `https://docs.aws.amazon.com/athena/latest/ug/vpc-flow-logs.html`

- The ability of CloudTrail to send trail events to CloudWatch Logs; more information is available at this URL: `https://docs.aws.amazon.com/awscloudtrail/latest/userguide/send-cloudtrail-events-to-cloudwatch-logs.html`

Going back to the AWS shared responsibility model, you are responsible for the premises where Outposts live and also capacity management. The AWS services and processes described in this chapter are valuable tools to help you improve your system availability and business continuity.

Amazon CloudWatch and AWS CloudTrail are cornerstones to implementing comprehensive management and governance of your hybrid architecture. Explore other services such as the following:

- AWS Config: `https://aws.amazon.com/config/`

- AWS Systems Manager: `https://aws.amazon.com/systems-manager/`

- AWS Service Catalog: `https://aws.amazon.com/servicecatalog/`

You can also expand your knowledge about management and governance at AWS; more information is available at this URL: `https://aws.amazon.com/products/management-and-governance/`.

Summary

In this chapter, we delved into the AWS capabilities at your disposal to monitor AWS Outposts. This chapter was heavily hands-on, with several examples to illustrate the potential of the services available to you. There is a lot more to it; Amazon CloudWatch is a very complex service, and it could have an entire book dedicated to it.

Congratulations on making it to this point; our journey has covered plenty of information and discoveries. It is not easy to get an AWS Outpost to play with, and I hope this book delivered on its promise to make you an Outposts hero. You already are one at this point, as the next chapters are light and easy by comparison.

It is now time to relax; we are on the last mile. In the next chapter, we will talk about what happens when things go wrong. That would be a nice title, but let's give it the name *Outposts Maintenance* for the sake of elegance. Onto the next chapter!

Part 3: Maintenance, Architecture References, and Additional Information

Concluding the journey, this part covers maintenance topics to ensure smooth and consistent operations, maximizing the value delivered by the product. Finally, it unveils resources available for customers to help design architectures and compelling solutions with AWS Outposts.

This part has the following chapters:

- *Chapter 7, Outposts Maintenance*
- *Chapter 8, Architecture References*

7
Outposts Maintenance

Monitoring your Outposts is a critical administrative task. Outposts resources are finite, especially if you are used to the *virtually unlimited* capacity of an AWS Region. Customers can make the mistake of not planning the AWS Outposts capacity and run into an `Insufficient Capacity` error whenever it runs out of capacity to provision resources. Additionally, service link interruptions, as well as S3 and EBS (Elastic Block Store) consumption, must be monitored.

If you have an established monitoring practice, you may be able to proactively react before something bad happens most of the time. Generally, you will be able to fix the cause of alarm yourself, for example, shutting down non-productive instances, deleting stale volumes, or deleting unnecessary S3 objects in a bucket.

However, as AWS CTO Werner Vogels said, "*Everything fails, all the time.*" It's an inconvenient truth that any physical hardware will inevitably degrade to a point of ceasing to function. If that ever happens to your Outposts, then it will require troubleshooting and maintenance, which is what this chapter is about. It will cover the following:

- Outposts hardware, software, and firmware maintenance
- Network troubleshooting checks

Hardware and software maintenance

Going back to the AWS Shared Responsibility Model, it was slightly modified for AWS Outposts. You are responsible for capacity management and, for this task, your procedures and practices used for monitoring Outposts will directly impact business continuity and application uptime.

However, the availability and security of the Outposts infrastructure, including the power supplies, servers, and networking equipment within the AWS Outposts rack, continues to be the responsibility of AWS, as well as the managed software components and AWS services that run on Outposts, such as the following:

- Firmware
- Virtualization hypervisor
- Storage systems

To fulfill this task, some form of monitoring is also required, and Outposts comes with all the necessary monitoring capabilities that allow AWS to collect telemetry and metrics from Outposts subsystems.

The hardware comes with sensors that monitor the health of various components, and the data points generated by these sensors are sent back to the AWS management network via the **service link**. AWS analyzes this data to identify hardware or software problems that could trigger maintenance procedures.

These maintenance procedures include placing degraded components into maintenance mode in preparation for physical replacement as well as patching firmware and hypervisor components on a regular basis.

These operations do not normally impact running instances on your Outpost, but there is always a small chance that a reboot of the Outposts equipment is needed to install the update. In such cases, AWS will send you an instance retirement notice for any instances running on that capacity. More information about instance retirement can be found at this URL: `https://docs.aws.amazon.com/AWSEC2/latest/UserGuide/instance-retirement.html`.

Software updates are automated and follow the same schedule and procedures defined by AWS for the infrastructure in the Outposts' anchor Availability Zone. All API calls made to upload data and configure services on AWS Outposts will traverse the service link in order to execute the requested action.

Software and firmware update activities depend on the service link to be carried out remotely. However, there might be times when an on-site presence is necessary. Let's remember, as a managed service, everything related to Outposts maintenance will be executed by AWS personnel, never third parties or contractors.

Let's also remember that AWS Outposts racks are shipped with security tags and tamper seals to provide additional assurance that they were not breached or violated. AWS performs checks on these elements before shipping Outposts to customer locations, and they must still be intact whenever AWS personnel arrive on site to perform post-installation and maintenance actions.

The only constant in life is change, and this is an inescapable fact. When your Outposts racks are deployed, several checks are performed, contacting facility providers to make sure AWS technicians could carry out their duties and follow AWS procedures setting up the rack.

Have you ever considered that these facilities might have changed their procedures and that something previously allowed can be forbidden in the future? Moreover, something that was not necessary to set up the rack but is necessary for a maintenance procedure may not be permitted. You must regularly check with your facilities provider that all necessary requirements for AWS technicians to carry out their jobs are allowed to enter the facility where Outposts rack is installed.

For example, AWS technicians must enter the Outposts' operating address with an Amazon-owned laptop and a network switch and take them to the rack position where the Outposts solution is installed. They also must leave the installation site with this equipment in hand. Make sure this is allowed at all times for your Outposts rack.

Another unforeseen possibility with facility providers is the requirement to destroy decommissioned parts or perform data degaussing on hardware. AWS does not allow these actions and its personnel must leave the site with all Outposts' non-functional or damaged components. If colocation or facility providers have strict policies about destroying and degaussing hardware, AWS Outposts is not a suitable choice for that location.

To ensure data protection, AWS is responsible for removing the **Nitro Security Key** (**NSK**) from a droplet, an action that renders data stored on the droplet unusable. It is the responsibility of AWS to destroy the data by turning the screw on the NSK that physically destroys the **integrated circuit** (**IC**) that stores the NSK's key material.

Physically destroying the NSK IC meets NIST 800-88 Data Sanitization requirements to **cryptographically erase** (**CE**) the data stored on the droplet. You can find more information about this guideline at this URL: https://csrc.nist.gov/publications/detail/sp/800-88/rev-1/final.

AWS would be incredibly happy to count on your help to maintain high levels of service uptime. That includes avoiding mobilizing AWS teams as a result of false alarms triggered by a scheduled event. If you are aware of a power loss or network maintenance in your facility that may affect your AWS Outposts, create a support case notifying AWS about the event.

This will greatly help AWS to be cognizant of alarms triggered from your Outposts; the telemetry could be correlated to maintenance events and not hardware malfunction that could potentially result in hardware replacement. AWS operates in partnership with the customer; if all parties mutually help each other, it creates a virtuous circle for mutual benefit.

It is human nature to plan for growth and success, not for failure. For a long time, IT followed the same path, focusing on deploying more applications and systems and adding more features, while leaving the discussions about outages, failures, and mishaps to some time in the future, hoping that such things would never happen.

Turns out these things happen, more often than we would wish for, and it is always best to take preventive actions than corrective actions. You must ensure that everything AWS requires to be able to maintain your Outposts is available during the whole duration of your contract term.

Take the maintenance of your Outposts as seriously as its operation; do not fall for the idea that you will never fall victim to a hardware failure because the risk is very low. Also, do not be just a spectator because most of the maintenance burden falls under AWS's responsibility. Be proactive and collaborative; success is a collective effort.

In the next section, let us explore some things that could go wrong with the network and give some tips on how to troubleshoot to narrow it down to the potential root cause. Troubleshooting is a collective effort as well; in certain cases, it might be you and AWS working together to pinpoint the faulty piece.

Network troubleshooting

The network substrate is a critical component of any IT architecture. Whenever your design requires a certain condition, the optimal performance depends on that condition being fulfilled. AWS Outposts has stringent networking requirements, and only only smooth network operations will allow the service to operate with peak performance.

The service link is a critical function and must be thoroughly monitored. If you receive an alert that your link is down, there are various actions that you can carry out to expedite the troubleshooting process because the problem's root cause might be on your end. To explore potential sources or problems, let's use as reference *Figure 7.1*, which depicts Outposts networking logical components:

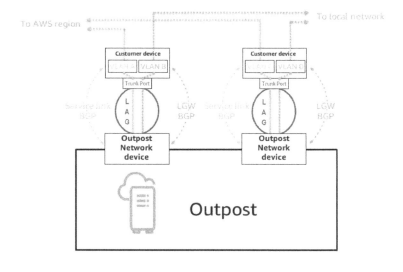

Figure 7.1 – AWS Outposts network logical diagram

Here is a checklist of potential causes of a **LINK DOWN** alarm on your end.

Instability or failure of your customer's device

There are times when your hardware may not completely fail, but discrete components can eventually malfunction. That may lead to packet losses, network flapping, random outages, or erratic operations. Overheating can occur if switch fans fail or data center cooling is insufficient.

Troubleshooting procedures in such cases must be done by accessing your device console and analyzing device logs. Make sure you collect extensive logs of your network devices with relevant data points; eventually, you may need to increase the log level to identify errors.

Switching devices configured with virtual stacking must be evaluated to make sure there is no instability or faulty components in the data backplane supporting the stacking protocol. Modular switches require an extra level of troubleshooting because problems can happen on the switch chassis or individual line cards.

Instability of your link aggregation (LAG)

AWS Outposts requires that your upstream device supports the **802.1Q** (**Dot1q**) standard, which implements the VLAN support needed to segregate traffic, and leverages **Link Aggregation Control Protocol** (**LACP**) aggregations with the corresponding **customer network device** (**CND**), which must comply with the IEEE 802.3ad standard.

The link aggregation capability in network switches goes by many names across the vendors. You can find documentation that refers to this concept with terms such as **channeling**, **bundling**, **bonding**, **teaming**, or **trunking**. The stability of the aggregation depends on individual member ports having the same configuration and operating at the same speed. Switch ports configured as auto-negotiation may change the port speed and cause instability, as well as inconsistent half/full duplex configurations.

Misconfiguration of your customer's device

Network topologies eventually undergo changes and these changes may affect your connectivity. Make sure changes don't cause adverse effects on BGP routing or break the BGP session peering. You may experience symptoms in only one of the logical channels, either the **local gateway** (**LGW**) or the service link. Or worse, you can have problems with both simultaneously.

If you suspect BGP, begin trying to ping the IP address of your Outposts network devices from your customer devices. If you don't get replies, check whether the logical interfaces are with the status UP. Logical interfaces may have a dependency on the link aggregation, therefore it must be stable and functional. Flapping interfaces cause fluctuations or complete interruptions in connectivity.

Make sure nothing changed with the BGP configurations defined for AWS Outposts during the network validation phase, such as **Autonomous System Numbers** (**ASNs**) or interface IP addresses.

Make sure the VLANs defined for the interfaces exist and are up and running. Other potential sources of problems are changes in route propagation, changes in BGP filtering, and changes in firewalls.

Faulty cables or transceivers

Wires and connectors can cause fluctuations in connectivity or a full disconnect. We consider the probability of a working fiber strand and optical connector to suddenly cease functioning to be incredibly low, and troubleshooting these elements is often the last action we take if they were operating normally for a long time.

But that can happen, of course. Fluctuations can be caused by inferior quality splices, accumulated dust, humidity, or bad connections in other segments. Dead connections may be related to fiber cables accidentally twisted, bent, crushed, or stressed by traction. Other causes for complete interruption of the light signal are contaminated optical transceivers, faulty splices, and faulty connections in the patch panel.

Problems with ISPs and WAN providers

AWS Outposts has a demarcation point, which is the patch panel for optic connections. You are responsible for connectivity beyond that element up to AWS Outposts service endpoints. Whatever happens, impacting connectivity from the demarcation point onward must be addressed by your teams.

The network that exists beyond your L2 switches and L3 routers can also be the reason for a network outage. Make sure any ISP providing WAN connections to you is operating normally and there is nothing preventing data traffic from Outposts from being forwarded to the service link or your local networks. Remember, the problem can also be related to returning traffic dropped or blocked at some segment of your network.

AWS provides a checklist for troubleshooting network problems at this URL: `https://docs.aws.amazon.com/outposts/latest/userguide/network-troubleshoot.html`. If you are using **Direct Connect**, work with your delivery partner to ensure that the services they provide are operating normally and, if not, implement any contingency measures you developed for events like these.

If you have manual processes to activate contingency plans, you must decide when to carry out the failover procedures. Rely on the service restore timeframe assessed by your connectivity provider before carrying out the failback. Notify AWS by opening a support ticket describing the situation and the expected timeframe to resume normal operations.

Summary

A widely known quote states that you cannot manage what you cannot measure. AWS Monitoring measures the telemetry and metrics from your AWS Outposts to proactively detect deteriorating hardware and service degradation. However, there are times when things break without warning, and you must also be prepared for this eventuality.

In this chapter, we highlighted that you also have responsibilities related to maintaining your Outposts service as fully operational. As Outposts is a managed service, sometimes we may fall into the comfort zone and think it is AWS's responsibility end to end to keep the service up and running. That's true for the most part, but you must never forget your duties as described in the AWS **Shared Responsibility model**.

Power, cooling, and network connection are your responsibility; therefore, you are in charge of the proper operation of these services. This chapter discussed common sources of problems, but of course, there are many more. Make sure you or your provider of choice operates your data center at a professional level to be able to benefit from Outposts' potential in full.

The finish line is in sight. You are now an AWS Outposts hero and this is no small feat. Hybrid Edge and related technologies are emerging topics, and being knowledgeable about AWS Outposts is a fantastic asset in your arsenal. In the next chapter, you will understand how you can expand your proficiency in AWS Outposts.

8
Architecture References

AWS Outposts is a complex service – unlike any other AWS service, it requires a lot of information from the customer to be able to order, activate, and operate the rack. However, the potential unlocked by AWS Outposts on the customer side is unmatched by any other similar offering on the market, bringing the power of AWS services to run on your data center.

Building solutions with AWS Outposts can be a challenging task. Even if you are a seasoned builder with AWS, you are likely to be bound to the comforts offered by AWS Regions where all services are available, and are not concerned about the underlying infrastructure and resource availability.

When you expand the scope of your architecture to interoperate with on-premises resources extensively, new skills are required and you will now have to consider limitations and failure scenarios that don't exist for AWS Regions. Likewise, performance will be dependent on factors extraneous to the cloud provider and they must be factored into your architecture.

AWS Outposts is a relatively new service with limited literature publicly available. However, AWS has an extensive collection of valuable resources to enrich your experience beyond the knowledge acquired in this book. This chapter will help you find various artifacts and resources on AWS Outposts:

- How to find resources on the Outposts product page
- How to find resources in the AWS Architecture Center
- Relevant blogs and white papers on AWS websites

How to find resources on the Outposts product page

Let's begin with the resources available on the AWS product site. It is easy to overlook these resources when you are beginning with AWS Outposts and trying to understand the product and fitment in the IT landscape. Head on over to the AWS Outposts product page at the following URL: `https://aws.amazon.com/outposts`. At the time of writing, the page looks as shown in *Figure 8.1* and you can identify distinct menus for **AWS Outposts rack** and **AWS Output servers** in the menu bar:

Figure 8.1 – The AWS Outposts service landing page

Select **AWS Outposts rack** and locate **Resources**. If you hover over this menu, there are two sub-menus: **Getting started** will give you very concise information about selection, ordering, launch, support, and maintenance. **Resources** will take you to advanced information categorized into the following sections:

- **Whitepapers**
- **Solution briefs**
- **Blogs**
- **Videos**
- **Training**
- **Infographics**
- **eBooks**

Figure 8.2 shows the menu dropdown:

Figure 8.2 – The Resources menu dropdown

Figure 8.3 shows the **Resources** page:

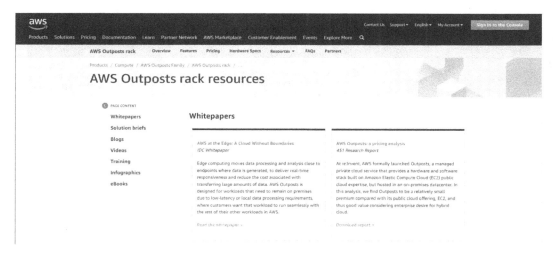

Figure 8.3 – Advanced-level Outposts rack resources

How to find resources in the AWS Architecture Center

Another excellent AWS resource is the **AWS Architecture Center** provision, which is available at this URL: `https://aws.amazon.com/architecture/`. This is a mandatory resource for AWS architects. To find the resources available for Outposts, you will need to perform a search. *Figure 8.4* shows the landing page and search box:

Figure 8.4 – The AWS Architecture Center search box

If you type Outposts into the search box, you will be presented with the reference architectures for AWS Outposts. *Figure 8.5* shows the results for this search criterion:

Figure 8.5 – AWS Architecture Center search results

Relevant blogs and white papers on AWS websites

White papers and blogs help to expand your knowledge greatly by providing in-depth technical content. Your one-stop shop for searching for white papers, as well as AWS guides, is found at this URL: `https://aws.amazon.com/whitepapers`. The AWS blog can be searched by accessing the landing page at `https://aws.amazon.com/blogs`.

Let us discover some key artifacts that will enhance your toolbelt to create hybrid architectures and will take your proficiency in elaborating on Outposts to the next level. Whenever you are talking to the customer's application owners, facility owners, or network owners, the content of these documents will be the north star for discussions and will certainly pave the way for a smooth and top-notch experience with AWS Outposts.

AWS Outposts High Availability Design and Architecture Considerations, a must-read, can found at this URL: `https://docs.aws.amazon.com/whitepapers/latest/aws-outposts-high-availability-design/aws-outposts-high-availability-design.html`.

This resource provides a thorough explanation of your options for implementing high availability and fault tolerance when creating architectures using Outposts. It breaks down the failure modes into five domains: the network, instances, compute and storage servers, racks and data centers, and AWS Availability Zones or Regions.

Next, we have a two-part blog about best practices for AWS Outposts in a multi-account environment. Part 1 can be found at this URL: `https://aws.amazon.com/blogs/mt/best-practices-aws-outposts-in-a-multi-account-aws-environment-part-1/`. Part 2 is at this URL: `https://aws.amazon.com/blogs/mt/best-practices-aws-outposts-in-a-multi-account-aws-environment-part-2/`.

These resources provide extensive guidance and information about sharing AWS Outposts resources, designing your AWS Organization, and utilizing AWS **Resource Access Manager** (**RAM**).

Details on the deployment of Amazon RDS on AWS Outposts with Multi-AZ availability can be found at this URL: `https://aws.amazon.com/blogs/database/deploy-amazon-rds-on-aws-outposts-with-multi-az-high-availability/`.

Multi-AZ RDS deployment on AWS Outposts was a feature long-awaited by customers and this document explains in detail how to set up replication via a local network between two racks, one running the primary DB and the other running the secondary DB.

A blog post about AWS Outposts instance management can be found at this URL: `https://aws.amazon.com/blogs/compute/managing-and-securing-aws-outposts-instances-using-aws-systems-manager-amazon-inspector-and-amazon-guardduty/`.

Security is a priority in any architecture and AWS offers services such as AWS Systems Manager, Amazon Inspector, and Amazon GuardDuty to enhance your security posture. This content provides a walkthrough of setting up these services to manage the instances running on Outposts.

Containers are a hot topic, application modernization is a strong trend, and AWS Outposts plays a key role in the AWS portfolio in this realm. A blog post about ECS on Outposts can be found at this URL: `https://aws.amazon.com/blogs/containers/amazon-ecs-on-aws-outposts/`.

This extensive content shows you how to deploy a solution based on ECS on AWS Outposts with step-by-step instructions. For EKS, we have another blog post written in a similar fashion, this time with walkthrough to deploy Amazon EKS on AWS Outposts using Terraform. It can be found at this URL: `https://aws.amazon.com/blogs/compute/building-modern-applications-with-amazon-eks-on-amazon-outposts/`.

Concluding our blog list, a blog post about migrating on-premises workloads to AWS Outposts can be found at this URL: `https://aws.amazon.com/blogs/storage/migrate-on-premises-workloads-to-aws-outposts-cloudendure-migration/`.

This content explains how to configure CloudEndure to replicate a source workload to an EC2 instance inside Outposts, test the target instance functionality, and finally, perform a cutover.

Lastly, let's refer to two sources of valuable information, not only about Outposts but AWS in general – first, AWS's official channel on YouTube, which can be found at `https://www.youtube.com/c/amazonwebservices`. Just search for Outposts and there will be no shortage of videos returned by this query.

Second, AWS Samples on GitHub. Check the link (`https://github.com/aws-samples`) and search for Outposts repositories.

This chapter unveiled some gems scattered across the huge assortment of content distributed across AWS websites. You would eventually stumble across some of them using web searches, but this collection provides point-and-click references to very important topics that will be a lot of fun to read for tech-savvy folks.

Summary

You are done – congratulations on crossing the finish line! Hopefully, you are exultant about your choice to become an AWS Outposts hero. There is a myriad of possibilities in the Hybrid Edge space, which awaits professionals with proficiency, and you are now capable of leveraging Outposts as a potential solution for your hybrid architectures with confidence and amplitude.

I would like to thank you immensely for enduring this journey and choosing this book. AWS Outposts is *the* service for me – the opportunity to sneak into the AWS data center infrastructure, pluck out a piece of it, and put it to work on our home turf.

The AWS portfolio is incredibly vast and fascinating, but whenever you are building architectures with AWS services in a region, there is physical hardware powering it up and that is comprised of AWS Outposts relatives. One distinctive objective of this book is to spark the flame of enthusiasm about AWS Outposts and what it represents to ignite its widespread adoption.

Even if you are a business that uses solely traditional commodity hardware from well-established vendors, at some point, you will need a greater capacity or face a hardware refresh. When that day comes, think Outposts! Hopefully, we made abundantly clear how distinct AWS Outposts is from what you are using today, and that is the perfect opportunity to begin your journey to the cloud – move to Outposts first and from there, the cloud and its full potential is one *service link* away from you!

There's more to come for the hybrid edge space – stay tuned and never stop learning. It was very satisfying to deliver this content and I will be thrilled to meet you in another learning session or live opportunity! Paraphrasing the famous TV series "*Now Go Build*", starring Werner Vogels, let's close by saying, "*Now Go Build with AWS Outposts.*"

Index

Packt.com

Subscribe to our online digital library for full access to over 7,000 books and videos, as well as industry leading tools to help you plan your personal development and advance your career. For more information, please visit our website.

Why subscribe?

- Spend less time learning and more time coding with practical eBooks and Videos from over 4,000 industry professionals
- Improve your learning with Skill Plans built especially for you
- Get a free eBook or video every month
- Fully searchable for easy access to vital information
- Copy and paste, print, and bookmark content

Did you know that Packt offers eBook versions of every book published, with PDF and ePub files available? You can upgrade to the eBook version at packt.com and as a print book customer, you are entitled to a discount on the eBook copy. Get in touch with us at customercare@packtpub.com for more details.

At www.packt.com, you can also read a collection of free technical articles, sign up for a range of free newsletters, and receive exclusive discounts and offers on Packt books and eBooks.

Other Books You May Enjoy

If you enjoyed this book, you may be interested in these other books by Packt:

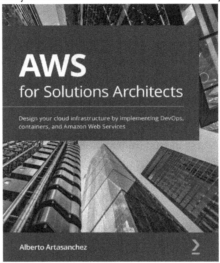

AWS for Solutions Architects

Alberto Artasanchez

ISBN: 9781789539233

- Rationalize the selection of AWS as the right cloud provider for your organization
- Choose the most appropriate service from AWS for a particular use case or project
- Implement change and operations management
- Find out the right resource type and size to balance performance and efficiency
- Discover how to mitigate risk and enforce security, authentication, and authorization
- Identify common business scenarios and select the right reference architectures for them

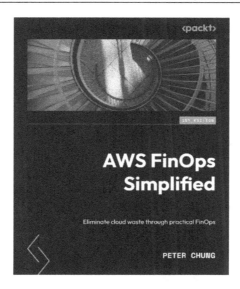

AWS FinOps Simplified

Peter Chung

ISBN: 9781803247236

- Use AWS services to monitor and govern your cost, usage, and spend
- Implement automation to streamline cost optimization operations
- Design the best architecture that fits your workload and optimizes on data transfer
- Optimize costs by maximizing efficiency with elasticity strategies
- Implement cost optimization levers to save on compute and storage costs
- Bring value to your organization by identifying strategies to create and govern cost metrics

Packt is searching for authors like you

If you're interested in becoming an author for Packt, please visit authors.packtpub.com and apply today. We have worked with thousands of developers and tech professionals, just like you, to help them share their insight with the global tech community. You can make a general application, apply for a specific hot topic that we are recruiting an author for, or submit your own idea.

Share Your Thoughts

Now you've finished *Simplifying Hybrid Cloud Adoption with AWS*, we'd love to hear your thoughts! Scan the QR code below to go straight to the Amazon review page for this book and share your feedback or leave a review on the site that you purchased it from.

https://packt.link/r/1803231750

Your review is important to us and the tech community and will help us make sure we're delivering excellent quality content.

Download a free PDF copy of this book

Thanks for purchasing this book!

Do you like to read on the go but are unable to carry your print books everywhere?

Is your eBook purchase not compatible with the device of your choice?

Don't worry, now with every Packt book you get a DRM-free PDF version of that book at no cost.

Read anywhere, any place, on any device. Search, copy, and paste code from your favorite technical books directly into your application.

The perks don't stop there, you can get exclusive access to discounts, newsletters, and great free content in your inbox daily

Follow these simple steps to get the benefits:

1. Scan the QR code or visit the link below

https://packt.link/free-ebook/9781803231754

2. Submit your proof of purchase
3. That's it! We'll send your free PDF and other benefits to your email directly

www.ingramcontent.com/pod-product-compliance
Lightning Source LLC
Chambersburg PA
CBHW060540060326
40690CB00017B/3561